D1209595

**Marketing
High Technology**

Marketing High Technology

William L. Shanklin
Kent State University

John K. Ryans, Jr.
Kent State University

LexingtonBooks
D.C. Heath and Company
Lexington, Massachusetts
Toronto

Library of Congress Cataloging in Publication Data

Shanklin, William L.
Marketing high technology.

Includes index.
1. High technology—Marketing. 2. High technology industries.
I. Ryans, John K. II. Title.
HC79.T4S49 1984 621.381'7'0688 83-48064
ISBN 0-669-06914-0

Third printing, June 1985

Published simultaneously in Canada

Printed in the United States of America on acid-free paper

International Standard Book Number: 0-669-06914-0

Library of Congress Catalog Card Number: 83-48064

To Joan, Andrea, and Courtney

To Cinda

Contents

Vignettes

Figures and Tables

Figures

Tables

Change Is the Only Constant

Introduction

Our ancestors were undoubtedly amazed by the revolutionary scientific devices and discoveries of their times that we take for granted today. Who marvels now over the electric light bulb, the automobile, or the Salk vaccine? Likewise, our children's children will hardly be impressed with present-day technological breakthroughs. The space capsule that first took astronauts to the moon will be but one of the many "Model Ts" of tomorrow.

Consequently, the *high* in high technology depends on the times. To early humans, the discovery of the wheel and its possibilities was advanced applied science. And such incremental gains in applied science, meager though they may later seem to be, are enormously important to us: Technology brings freedom to our lives from laborious tasks, facilitates our education through improved worldwide communications, and mitigates our worries over disease. Inarguably, technological progress also carries with it great responsibility, especially with regard to controlling technology's potential for destructive use. Advances in weaponry systems, exporting computer capabilities to totalitarian regimes for possible use in surveillance of dissident citizens, and genetic engineering have all raised moral and ethical issues in our day. There are essential economic ramifications from high technology as well. These are the focal point of this book.

The nation that thrives in high technology is going to have a superior material standard of living vis-à-vis its sister nations and a larger say in the conduct of world affairs. Power and influence inevitably flow to the countries most adept at fostering the discovery and advancement of technology. (The same can be said for the individual states within the United States.) And adeptness depends not only on the collective ability of a nation's scientists to make headway in the laboratory, but also on the prowess of its business people in developing and marketing incipient technologies in a commercially successful way. Indeed, the development of high technology that fulfills a significant human need or desire, present or future, is the critical marketing task. Keen and perceptive need discernment and fulfillment do not make the actual selling job superfluous, but they certainly make it easier.

Technological innovation in the United States and elsewhere in the free world has often been the product of individual free-lance inventors who saw their ideas through to ultimate success, frequently in the face of many trials and tribulations and with great odds against the inventor. Thomas Edison's advocacy and use of the industrial research laboratory in the nineteenth century began to change this pattern and to shift technological innovation to corporate sponsorship—although one-person entrepreneurs did and still do occasionally succeed.

With increased corporate sponsorship of technological innovation came an important dichotomy between those people who conceptualized and invented technology and those individuals who developed, financed, and marketed it. Risk of failure was also spread over more individuals, primarily the corporate shareholders or oftentimes the venture capitalists. We feel that this distinctive and critical division of labor between the technological and business sides of innovation has been largely glossed over in the ongoing debate in the United States about the causes of and remedies for the nation's eroding technological leadership.

As the United States has watched its technological edge erode in various fields, we have heard it said repeatedly that what the United States needs is more scientists, mathematicians, and engineers, as well as improved schooling in the sciences at all educational levels. We could not agree more with these recommendations, as long as they are presented as only one part of the solution and not a panacea. We believe that even if the United States somehow could miraculously produce enough capable scientists, mathematicians, and engineers overnight, our technological concerns as a nation still would not be resolved. The most brilliant scientist or engineer might be a poor discerner of market needs—the place where every technological breakthrough must ultimately prove its worth. Scientists and engineers are just as susceptible as the rest of us to becoming enamored with their own work and consequently to making errors of judgment about how others will perceive and accept it. Technologically oriented corporate employees need solid direction from the marketplace of today or tomorrow if their efforts in the laboratory are to coincide with corporate goals and strategies. The role of marketing is to provide this direction. As Dr. Edward W. Ungar of Battelle Memorial Institute told us, "In corporate R&D, most ideas for new products need to be evaluated against the test of whether or not the product will be accepted in the competitive marketplace."

We began the project which resulted in this book by first testing our intuition that the same marketing concepts and techniques that are so largely responsible for the high standard of living enjoyed by mass distribution societies need adaptation and sometimes refinement to fit the specialized and often singular needs of companies engaged in high technology. Our belief was based on our discussions with a number of executives in high-tech companies. We also asked ourselves two questions: "Even if this is true—that traditional marketing concepts and practices need adaptation and refinement—is this topic worth researching and writing a book about? Would we really make a contribution?"

After further discussions with management in a diversity of high-tech companies and with government officials concerned about U.S. technological leadership and the U.S. export position, our answers to these questions were resoundingly affirmative. To say that many of the people we asked were excited and encouraging about a research-supported practical book on marketing high technology would not be an overstatement. In our opinion, the demand for a book on marketing high technology only rarely resulted from a situation where engineers and scientists were running a company without the counsel of personnel with first-rate marketing knowledge and savvy. Something far more fundamental was at work. We concluded that conventional marketing concepts and methods work best once a technology has matured and established itself in the marketplace. But in the initial stages of market acceptance—what we term the supply-side stages—the same marketing concepts and techniques are sometimes not applicable. In this book, we look at both supply-side and demand-side marketing and suggest how each can be employed effectively. In particular, the volatility of high-tech industries calls for creative approaches to product management and to human resource staffing requirements (entrepreneurial managers versus maintenance managers) over the product and technology life cycles. Similarly, the intrinsic complexity of high technology has interesting and meaningful implications for the communications functions—advertising and personal selling.

Since we started the project which led to this book, our judgment about the need for practical research-based guidance on the marketing and management of high technology (supply-side marketing, we call it) has been reaffirmed by the people we have come across in our research and by the number of leading universities that have established programs within their business schools on entrepreneurship and the management of innovation. Most of these programs are partially funded by high-technology companies of note, and at least one major university has begun to offer a joint graduate degree between its schools of business and engineering.

After we satisfied ourselves that there was a real need for a book on marketing high technology, we turned to collecting data. Initially, we searched for existing marketing literature that pertained precisely to marketing high technology. But we were not content merely to rely on an individual author's claim for the usefulness of a concept or technique; we wanted to see for ourselves what the experience had been, if any, with a given concept or technique under actual high-tech marketplace conditions. To do so, we solicited the opinions of knowledgeable executives within successful high-tech firms. We wanted to know, "What has been the track record for . . . ?" In those instances where the marketing literature appeared to be silent on the usefulness to high technology, of a marketing concept, method, or activity, we proceeded in a like manner. Further, we sought new ideas and explored possibilities we had developed. Our efforts were comprehensive—we covered virtually every fundamental area of marketing management and research.

Our data came from executives in a wide and representative cross-section of high-tech industries, from ventures as diverse as robotics and biotechnology. The data were collected over several years by means of in-person and telephone sessions, written questionnaires, and a number of in-depth studies on specific marketing areas, such as personal selling and advertising in high-technology markets. Our data sources have one thing in common beyond being in high technology: They are all proven winners in the high-tech sweepstakes. While much can be learned from the mistakes of companies that have failed (and we cite and consider such mistakes from time to time), we hardly want to offer advice and counsel on marketing high technology that is founded on the experiences of losers in the marketplace. Consequently, when we refer to what high-tech firms do in the way of marketing, we mean successful companies that have proven themselves in marketplace competition over enough years to demonstrate that their success is not a fluke.

Let us say at the outset that this is not a book on how to market high technology. Rather, it is a book on what to do to market high technology in order to increase the chances of success—what marketing concepts, techniques, and activities to use (or not to use). The details and intricacies of how to are left to other sources.

Acknowledgments

The development of any book requires the advice and assistance of many individuals. Writing a book in an area that has not been previously explored makes such help even more essential. While we totally accept the responsibility for the material presented here, we wish to express our gratitude to the many persons who have provided us with valuable perspectives.

Our particular thanks go to Michael R. McGinley of Corning Medical and Scientific, Corning Glass Works; Ronald Evenson of Mill-Rose Laboratories, Inc.; Timothy J. Bublick and Nancy Lieber of The DeVilbiss Company; Robert Donath (*Business Marketing*); William Kolarik (U.S. Department of Commerce); Richard Reiser of Reiser Williams DeYong; Erik Borre of Novo Industri S/A; Robert Pacenta of Hewlett-Packard; Dean Peebles of Goodyear International; Dr. Alan Steggles of the N.E. Ohio College of Medicine; Stephen N. Bowen of NCR; Dr. Richard T. Hise of Texas A&M; Dr. Raj Aggarwal of the University of Toledo; and Mark G. Frantz of Frantz Medical Development Ltd. for their contribution at one or more critical points in the process. Further, we wish to acknowledge those many high-tech company executives who took the time and made the effort to respond to our detailed surveys and follow-up inquiries; without their assistance this book would not have the type of strong factual base that we desired.

At the operational level, we wish to gratefully acknowledge the efforts of Janet Currey, our typist; Janet Parkinson, our editor; Lori Mitchell, research assistance, of the University of Toledo; Susan DeSai, research assistant, of Kent State University; Kem Mraovich of Kent State University; and our secretaries, Susan Andrews of the University of Toledo; Mary Bruce of Kent State University; and Bernedette Messina of Kent State University.

1 Technological Innovation and Marketing

Certain imperatives define our mission: creating a technological advantage—that is, research; and creating a commercial advantage—or marketing.

Technological progress in any society is fostered and facilitated by the presence of three key ingredients—innovative and creative people, an economic climate conducive to entrepreneurship, and the managerial know-how both for stimulating new concepts and taking the promising ideas through to successful initial commercialization and beyond. These three elements complement one another and produce synergies. Innovation works best when all three ingredients are present. To see that this is so, one needs only to compare the technological track records of the capitalist-oriented, open-society nations of the world to those of the closed, socialist countries. Historically, in the technological-leader nations, there has been an exalted role for the uncommon person (in this case, the entrepreneur), whereas in the technological-follower countries of the world, it tends to be the common person who is idealized.

For innovation to thrive, technological pioneers, first and foremost, are needed to dream their dreams and follow them through to a glorious or an inglorious end. Success often comes to individuals who seem to possess the knack for snatching victory from the jaws of defeat—driven persons who won't take no for an answer. This description aptly fits the vast majority of the founding fathers of great modern-day fortunes and companies.

During some of the darkest days in U.S. economic history, Chester Carlson, a patent attorney, doggedly pursued the idea that culminated in the Xerox Corporation. He was forced to set up shop in his kitchen because the leading mimeograph firms spurned Carlson's efforts to sell his then-unwanted idea of an electrostatic printing process.[1] Even in the Great Depression of the 1930s, there was a place for an incurable optimist with the temerity to challenge the conventional wisdom, the perseverence to keep

---

The epigraph is a remark by Robert P. Luciano, President and Chief Executive Officer, Shering-Plough Corporation, in his presentation before the New York Society of Security Analysts, June 17, 1983.

going in the face of adversity and rejection, and the tolerance to cope with the considerable risk of failure. With initial help from the Battelle Memorial Institute in Columbus, Ohio, now the world's foremost contract research and development organization, and later through participation by the Haloid Corporation, then a small Rochester-based manufacturer of photographic and photocopy papers, the entrepreneurial Xerox story took shape over two-and-one-half decades. In 1961, the Haloid Xerox Company changed its name to the Xerox Corporation, and the rest of the story is well known.

Carlson's story is an old one that is being relived today by others like him. In extolling various contemporary entrepreneurs, a popular advertisement reads, "Here's to gut feelings and those who still follow them." A motivational book on the market describes famous rejections that ultimately became rousing successes because the dreamers behind them would not quit. The underlying theme in all these rejection-turned-success stories is that the people involved had the strength and conviction to keep trying against great odds. Maybe it is the "animal spirit" that economist John Maynard Keynes thought was the root of all entrepreneurship. Theodore Roosevelt might well have been paying high tribute to the past, present, and future pioneers of technological advancement—and failure—when he said, "It is not the critic who counts; not the man who points out how the strong man stumbles, or where the doer of deeds could have done them better. The credit belongs to the man who is actually in the arena, whose face is marred by dust and sweat and blood; who strives valiantly, who errs and comes short time again and again . . . and who at worst, if he fails, at least fails while daring greatly . . ."

Chester Carlson's stick-to-itiveness and drive and the related traits described by Theodore Roosevelt are unusual, but far from unique, even in today's more sophisticated and complex times. And these traits are just as essential as always to productive innovation.

A second ingredient in the innovation mix is an environment that permits and encourages creativity. What about the U.S. environment, now and in the recent past? First, the results: The number of U.S. patents issued to Americans has fallen 25 percent since 1971, while the number issued to foreigners has increased 14 percent. Approximately 40 percent of the patents now issued by the United States government go to individuals from other countries. America's most notable foreign competitors, Japan and West Germany, outperform it in both capital investment and research and development as a percentage of total national output. The antidote to the problem has been described by the successful and innovative corporation, United Technologies: an economic atmosphere more encouraging to innovation. The company suggests the need for policies and programs to expand savings and investment, to enhance the climate for risk-taking and innovation, to encourage adequate corporate profits, and to ease the tax and

High-Tech Movers and Shakers: Yesterday and Today

A Legend in His Time

Thomas Alva Edison deserves the accolade "greatest inventor." The "Wizard of Menlo Park" (after the New Jersey location of his first laboratory) patented 1,093 inventions in the United States and 2,000 to 3,000 more in foreign countries. Prominent among these were the electric light bulb, the motion-picture projector, and the phonograph. Edison also improved both the telephone and typewriter. A personal guideline that served him well was "First, be sure a thing is wanted or needed, then go ahead."

Edison's many works were the legacies of a man who had but three months of formal schooling. Remarkably, early in his life he was thought to be intellectually slow, an experience later to be shared by a young Albert Einstein.

True to his beliefs that "genius is two percent inspiration and ninety-eight percent perspiration," Edison had great perseverance and patience. When working on a project, he would toil 18–19 hours a day for stretches of 5 and 6 days at a time. Before achieving a breakthrough on his storage battery, he conducted 8,000 tests without success. Edison's acute hearing problem no doubt aided him in giving his work undivided attention.

Thomas Edison was interested in financial rewards only because they made possible further innovation. He set up the first equipped industrial research laboratory (West Orange, New Jersey, 1887), thus establishing the all-important precedent for organized research.

An Entrepreneur's Entrepreneur

William R. Hambrecht can appropriately be described as an entrepreneur behind entrepreneurs. His venture capital firm, Hambrecht and Quist, finances mostly high-tech business endeavors in their start-up and formative stages. Hambrecht, a Princeton University graduate who says his worst subject was science, is known in high-tech circles as the "banker to the future." He has "incubated the future" by financing over 150 ventures, including, for example: ARGO Systems (innovator of advanced radar and signal reconnaissance military systems); Avantek (microwave components); Collagen Corporation (their product takes the plastic out of plastic surgery); DiaSonics (pioneers of diagnostic imaging alternatives to x-rays); and SofTech (programming applications for ultrasophisticated computer languages). Hambrecht and Quist has brought some sixty high-tech firms public, including the initial offering of Apple Computer stock.

Whenever William Hambrecht evaluates potential venture capital commitments, what he looks for initially is evidence of the entrepreneur. He says, "We back people, not products." As a rule of thumb for the venture capital business (the risky business of financing innovation), one out of ten investments is expected to have a very large return on investment within five years.[a]

[a]Biography of William R. Hambrecht adapted from "The Low-Profile Impresario of High Technology," *United—The Magazine of the Friendly Skies,* April 1983, pp. 118–117.

regulatory burdens that undercut firms' ability to create, compete, and produce.

The third component of technological innovation is managerial know-how to develop and market high-tech products. This book concentrates largely on this aspect of innovation. Innovation can and does occur through happenstance. History is replete with mostly fortuitous beginnings for technological innovations. However, the structure of U.S. industry is far different now from the way it was in the days of Edison and Ford. There is a real need now for much more planned technological innovation within the corporate structure, rather than having society rely on the single-person venture that has historically characterized technological breakthroughs. If the United States is to maintain technological leadership, it must be achieved primarily within the corporate framework. Technological research, development, and marketing simply require too much capital and expertise for the single-venture entrepreneur.

Fortunately, marketing thought and practice has advanced and been refined to the point where a great deal is known about enhancing the success of high-tech product endeavors. Marketing technology has kept pace with engineering technology. Yet, in high-tech firms, there is too frequently an absence of adequate marketing knowledge and skills; consequently, the product failure rate is higher than it should be or would be with the application of appropriate marketing concepts and methods. Marketing knowledge and expertise is vital to the success of high-tech firms. It is often easier to find and hire people capable of researching and engineering high-tech products than it is to find and employ people knowledgeable about how to mitigate the consumer nonacceptance risks inherent in high-tech product conceptualization, development, and commercialization. The difficulty lies not so much in inventing and engineering new products, but in the application of the technology in a way that will result in market acceptance and its all-important by-products, profitability and return on investment.

A Place for Supply-Side Marketing[2]

Marketing books have traditionally been demand-side oriented and have devoted most of their attention to consumer products. High-technology marketing, however, involves demand-side *and* supply-side marketing expertise and pertains to consumer and industrial products/processes.

Why is the supply-side so essential? We are concerned with unleashing and managing Yankee ingenuity, the integrating theme of which is supply-side marketing: the marketing of products that are on the cutting edge of technology; those that create or revolutionize markets and demand; the kinds of products which make possible significant advances in standards of living or new ways of doing things. In short, the types of products that earned for the United States the reputation for technological innovativeness in the first place.

The supply-side concept has its origins in classical, pre-Keynesian economic theory. The idea that supply can create its own demand is known as Say's Law, after the nineteenth century French economist Jean Baptiste Say. Until recently, this assertion had lost much of its intellectual currency, as most modern-day economists were educated on the familiar demand-side precepts of John Maynard Keynes.

Just as most Keynesians are in varying degrees uncomfortable with supply-side economics, individuals educated and reared on demand-side marketing (i.e., market demand *should* precede and trigger product development and the formulation and funding of marketing strategies) look skeptically at supply-side marketing thinking. To many of the demand-siders, supply-side thinking is dangerous, product-centered, and myopic—antithetical to the demand-side focus on consumers and their needs and wants. This point represents far more than mental fencing: the economic penalties of corporate top management not coming to grips with supply-side concepts can be very real and can result in missed opportunities.

Supply-side marketing, then, refers to any instance where a product can create a market—in other words, a demand for itself—instead of the conventional other-way-around. Or, put differently, the product is responsible for the demand, rather than the demand being responsible for the product. Supply-side marketing pervades many of the Chester Carlson-Xerox and kindred high-tech Horatio Alger success stories.

The essence of supply-side marketing has been delineated by Akio Morita, who is chairman and chief executive officer of Sony Corporation. In his view, the more innovative a product is, in terms of being a radical departure from the past, the more likely it is that potential purchasers might not initially see its usefulness or benefits. For instance, in 1950 Sony marketed a tape recorder. Even though it was a real technological breakthrough, the general public thought of it as a toy. The tape recorder was not seen as a device for learning languages or storing speeches. Experiences like this have led Morita to believe that, for technological innovations, markets must be created, not surveyed.[3] This philosophy is reflected in the slogan printed on Sony helicopters: "Sony: Research Makes the Difference." The prominence given to this theme emphasizes the degree of importance that Sony attaches to its chairman's philosophy.

Of course, what Morita is referring to are high-technology products, especially those that potential customers (industrial or ultimate consumers) cannot readily relate to present products and current life styles, such as automobiles in their early days, radios in the 1920s, television in the 1940s, and hand-held calculators in the 1960s. Biotechnics or gene-splicing is a more current example. Another is Sony's filmless still camera that electronically records images on magnetic discs for television viewing. Sony is developing a companion electrostatic printer that produces hard copies on signal from an adapter. If that's not enough, the revolutionary camera can be plugged into a videotape recorder to make movies; photos can be transferred into a videotape to form an album of pictures; and pictures can be transmitted instantly via telephone.[4] This is the type of product requiring supply-side marketing.

In each of these cases, it would have been extremely difficult for anyone to have conducted a survey or used any other kind of traditional quantitative marketing research to derive anything approaching valid demand estimates for making a go/no-go decision. Imagine attempting to obtain early consumer reaction to television by way of a concept test. If there were a prototype to view, it still would have been hard for most prospective consumers to contemplate and comprehend what it would be like to watch, say, Milton Berle performing his antics on a 10-inch screen every week. How useful would it have been to ask prospective consumers if they would buy such a contraption, even if the inquirer had a good estimate of what it would cost at retail?

In the field of biotechnology, consider the case of monoclonal antibodies. Monoclonal researchers have developed simple, fast, and inexpensive diagnostic tests for hepatitis, prostate cancer, pregnancy, venereal disease, and a number of similar conditions. These tests do not require sophisticated instrumentation. Yet, many of the early monoclonal antibody-based tests did not demonstrate readily apparent advantages over existing diagnostic methods. Consequently, the outlook for market success was poor. Based on these pessimistic results, ten years ago who could have predicted that monoclonal antibodies would have already created a major new medical market—and that an astronomically lucrative portion of that market is yet untapped?[5]

Much of what is written and said about how to develop products from the idea stage through to commercialization is, pure and simple, often inapplicable, sometimes erroneous, and usually in need of refinement when it comes to high technology. Traditional demand-side marketing research, for example, is of limited value, as Sony's Morita points out, since it requires those surveyed to have at least a rudimentary idea about, and a frame of reference for, what they are being asked. More in-depth techniques, primarily what are known as focus groups and concept tests, are better. The fact is, many of the successful technologies of today would not have been introduced had soundings of the then-current marketplace been heeded. Reacting, for example, to Sony's filmless still camera, a spokesman for Eastman-

Kodak says, "It will never have mass-market appeal." Maybe. In classic supply-side fashion, Sony's chairman boldly predicts otherwise: "This product, like the videotape recorder, will open a whole new market."[6]

Supply-side marketing is a more aggressive approach to business than its demand-side counterpart. It is founded on the entrepreneurial spirit, calculated but high risk-taking, and the bent to exploit the unknown for great personal and/or corporate gain. With sketchy and quite likely discouraging hard information about the present market demand or present consumer needs and preferences, supply-side marketers develop and introduce products based on intuition and judgment about future market demand, as well as on just plain fortitude. Many of them fail—that is to be expected—but a few also succeed. On the successes, great progress in technological standards of living and working is made.

Supply-Side Marketing for Every Organization?

In his preface to *Poor Richard's Almanack* in 1758, Benjamin Franklin wrote an essay titled "The Way to Wealth." He offered this eternal strategic verity: "Great estates may venture more, but little boats should keep near shore." The correctness of Franklin's risk-taking maxim as it applies to the modern business organization is abundantly evident.

In the area of external corporate growth, consider the many failed conglomerate mergers and acquisitions of the 1960s and the continuing wholesale redeployment of corporate resources to fewer basic businesses for each company.[7] Regarding internally generated corporate growth, the PIMS (Profit Impact of Marketing Strategy) project (located at the Strategic Planning Institute in Cambridge, Massachusetts) offers more corroboration. Using data from many businesses in a plethora of situations, PIMS analysis found that if a firm's market share is weak, high R&D spending generally depresses its return on investment. A high ratio of R&D to sales tends to be profitable for large market share companies, but definitely works to exacerbate ROI for low share firms.[8]

By its very character and definition, supply-side marketing is not for every organization. In almost all markets and industries, there will be (and according to Franklin and PIMS, should be) a preponderance of technological followers rather than leaders. And certainly many high-tech companies initially (and sometimes subsequently) play a follower role. The vast majority of markets and industries also simply do not warrant the classification as high technology. Thus, supply-side marketing will usually have relevance only to small portions of the product/division portfolios of even the largest commercial concerns. Yet, supply-side products and divisions are disproportionately important in auguring long-term corporate profitability and, more broadly, in technological progress for society.

Perils of Too Much Demand-Side Influence

The demand-side concept of marketing is generally held to contain certain tenets, among these being a profit requirement and the integration of marketing activities under a line marketing officer. The linchpin tenet is consumer orientation. The merits and shortcomings of this idea have been argued through the years. Nonetheless, the concept is now firmly entrenched in marketing thought and practice.

Irrespective of whether this idea is considered to be a theory, a hypothesis, philosophy, manifesto, or whatever, it is an incomplete one. It is not whole because it pertains only to demand-side marketing. Although future consumer needs and wants are unknown, the implication is that present consumer demand should determine whether to commit corporate resources to product development and marketing programs. That, of course, is as it should be in most instances. But this demand-side approach does not really accommodate supply-side marketing, where breakthrough high-technology products ultimately create their own demand and markets (i.e., future consumer needs and preferences).

What the demand-side concept provides is plenty of slight product modifications or differentiations (e.g., annual model changeovers and "new, improved" reformulations), typically large doses of market segmentation to meet existing needs more precisely, and a large measure of risk-reducing marketing research to guide these efforts.

The consulting firm of Booz, Allen, and Hamilton interviewed twenty-five major consumer product marketers, outstanding firms such as General Electric and Ralston Purina.[9] It was found, predictably, that most of these companies define new products from an external perspective; by how similar new products are to the existing ones and how consumers view them. Significantly, it was revealed that most companies do not vary their success criteria by product or strategic requirement, thus focusing attention on present markets and consumer needs and the short-term profit possibilities and lower risks they represent. Booz, Allen, and Hamilton commented that, although external considerations are important, the initial question addressed should be how a new product fits overall corporate strategy. For instance, what products will be coming on stream three, five, and ten years hence?

One supply-side advocate, David Smick, proposes that what is needed in U.S. companies is less of the cautious demand-side approach to business and more of what the economist Joseph Schumpeter labeled the creative destruction of capital—the process whereby a new idea comes into the mar-

How Top Executives in High-Technology Companies See Themselves—Their Behavioral Characteristics and Motivations

High-technology companies operate in a turbulent world on the cutting edge of industrial evolution. Changes that often take many decades in mature industries can and do occur virtually overnight in high technology. One scientific discovery or unique application of existing knowledge is all it takes to turn an entire industry upside down or even destroy it forever. Executives who work in this kind of milieu have been described as possessing such characteristics as indomitable will, animal spirit, and riverboat gambler instincts.

From our discussions with executives in high-tech firms and from our reading on human creativity and innovativeness, we were able to discern words and phrases that came up time and again in describing technological pioneers—people who, by their deeds, had manifested the mysterious entrepreneurial spirit. We took many of these words and phrases and developed the written measurement instrument shown below. Then, using the scale, we asked numerous presidents and chief marketing executives of high-technology companies to describe their own individual behavioral characteristics and motivations. (To see the actual behavioral/motivational composite, turn to page 201 in the appendix.)

In each of the following items there are two words or phrases that are polar opposites which are separated by a seven-space scale. Please look at each pair of words and then place a check mark (✓) on the scale in the place that you believe best describes you as an individual. Please record your first impressions.

Traditionalist	_ _ _ _ _ _ _	Avant Garde
Nonliteral Thinker	_ _ _ _ _ _ _	Literal Thinker
Nonlinear Thinker	_ _ _ _ _ _ _	Linear Thinker
Motivated by $	_ _ _ _ _ _ _	Not Motivated by $
Tortoise	_ _ _ _ _ _ _	Hare
Gambler	_ _ _ _ _ _ _	Risk Averse
Nonintuitive	_ _ _ _ _ _ _	Intuitive
Entrepreneurial	_ _ _ _ _ _ _	Nonentrepreneurial
Inferential Thinker	_ _ _ _ _ _ _	Noninferential Thinker
High Associative Skills	_ _ _ _ _ _ _	Low Associative Skills
Illogical	_ _ _ _ _ _ _	Logical
Analytical	_ _ _ _ _ _ _	Nonanalytical
Noncreative	_ _ _ _ _ _ _	Creative
Word/Symbol Oriented	_ _ _ _ _ _ _	Pattern Oriented
Nontechnology Buff	_ _ _ _ _ _ _	Technology Buff
Promote/Welcome Change	_ _ _ _ _ _ _	Resist Change
Extrapolative Thinker	_ _ _ _ _ _ _	Nonextrapolative Thinker
Low Ambiguity Tolerance	_ _ _ _ _ _ _	High Ambiguity Tolerance
Introvert	_ _ _ _ _ _ _	Extrovert

ketplace, making existing capital worthless. The largest American corporations—the so-called *Fortune* 500—have contributed almost zero net job growth for over a decade. Most *new* jobs have come from small and young firms (twenty or fewer employees and four years or less in business). Nearly 100 percent of new jobs in the northeastern United States come from these small firms and almost 70 percent nationwide. These firms know about and strive for creative destruction, while the more established companies have a vested interest in preserving the status quo and their own economic preeminence.[10]

Creative destruction and innovation have long been the hallmarks of U.S. industry. When one ponders the significant technological innovations of the past one-hundred years, it is indeed remarkable how many of them were conceived and developed in the United States. This history of success is no mere coincidence. The people of the United States have encouraged technological progress through direct and indirect aid to research, development, and education. In addition, they have allowed great economic rewards to accrue to successful entrepreneurs. Successful entrepreneurship can have immediate psychic rewards as well. Many noted entrepreneurs—Bell, Edison, McCormick, Ford, and others—were viewed as folk heroes in the annals of American history while they were still living.

More technological innovation of this kind can be forthcoming. Although there is a long way to go, federal government actions in the United States are beginning to promote research and development and capital formation. Notably, provisions for tax-deferred individual retirement accounts for *all* income-earning U.S. citizens and, if applicable, their nonworking spouses, meaningful tax cuts for individuals, better tax treatment for the savings- and investment-prone high-income citizen, a 25 percent credit for increasing industrial research, and indexing of tax brackets to net out the effects of inflation, are all steps that serve to increase the savings and investment necessary for healthy innovation. If it survives, the tax-bracket indexing provision of the Economic Recovery Act of 1981 may prove to be one of the most useful tax reforms for encouraging savings and investment ever installed in the tax code of the U.S. government.

Even with these advances, there is much yet to be achieved. Of all the nations of the world, the United States is in the forefront in providing tax incentives to consumption rather than savings and investment. Consider the U.S. tax provision that allows the taxpayer to treat interest expenses as an itemized deduction. This loophole provides a societal impetus to debt-financed consumption vis-a-vis savings and investment. What change can be made is debatable. With the possible exception of the politically explosive issue of deductible mortgage interest payments on primary residences, it is feasible for Congress to pass legislation to exclude interest payments as a tax-deductible item. A more remote possibility, but also a highly desirable one from the standpoint of boosting high-tech innovation, would be the institution of a value-added taxing system—a consumption-driven method of taxation that numerous European countries use to provide economic inducements to savings. (Compare these 1981 figures for gross fixed capital formation as a share of gross national product: Japan, 31 percent; Canada, 24.5 percent; West Germany, 22.9 percent; France, 21.2 percent; Italy, 20.3 percent; United States, 17.1 percent; and United Kingdom, 16.9 percent.)

Regulation and antitrust have been real barriers to innovation. But discussions and changes in the right direction are under way. However, the regrettable fact is that the United States had to be economically threatened and wounded before action was forthcoming. It was, for instance, only after the U.S. balance of trade over the years worsened that elected members of the federal government finally agreed upon the Export Trading Act of 1982 as a means to facilitate U.S. trade in foreign markets.

The chief executive officer of the widely respected Control Data Corporation, William Norris, has called for modifications in U.S. antitrust policy to allow for additional technological cooperation between and among U.S. companies.[11] Because these companies often needlessly duplicate research and development efforts, and thereby waste scarce resources, Norris sees technological cooperation as desirable, particularly at a time when U.S. technological leadership is being challenged vigorously by foreign competition. Control Data has itself ventured with a number of other large semiconductor and computer concerns to form the Microelectronics and Computer Technology Corporation for the express purpose of deriving technologies for use by the members. Significantly, the technologies will be licensed to other companies at reasonable fees—a valuable resource indeed, especially for smaller high-tech firms. Norris has called upon the federal government to encourage this type of cooperative venture, rather than only sometimes permitting it. Presently, both antitrust laws and Justice Department guidelines discourage instead of encourage cooperation among high-tech companies. Fortunately, progress is being made in Congress, where there is growing bipartisan sentiment to do just what the Control Data chairman advocates.

Productive technological cooperation will require some changes in corporate attitudes as well. Boards of directors and top management will have to look well beyond their own provincial interests. As Control Data's Norris points out, for the technological progress at the lowest cost to society, corporate statesmanship is the order of the day.

In addition to the brightening economic and regulatory climate in the United States and commendable efforts like those of Control Data to develop and share technology, the heretofore mostly demand-side perspective in corporate America is starting to yield to accommodate more of a supply-side orientation. The old idea that exhausting the seed corn this season means bad times next season is taking hold. The *Wall Street Journal* has observed that companies are increasingly viewing their technology supply in the same way that oil companies look at their reserves and are recognizing that new technological discoveries are essential if profits are to keep coming over time.[12]

That is the bright side. After years of neglect, many firms are increasing the R&D expenditures that will inevitably result in technological progress. In another study by the consulting firm of Booz, Allen, and Hamilton, this one a technological survey of executives in twelve of the largest U.S. companies, all respondents agreed that new technology is vital to productivity growth and to new market development. The managers appear to be willing to sacrifice higher earnings now to finance big investments in promising technology. They are inclined to subjugate at least some current demand and profit considerations to longer term supply-side strategic concerns. Yet, a strong caveat was added—few of the companies have the faintest idea of how to manage their technology efforts effectively. The really difficult part of technological innovation is not the hiring of scientists and engineers; what is arduous is managing a firm's R&D efforts so they are consistent with the company's business objectives.[13] According to the president of Carnegie-Mellon University, Richard Cyert, *America's most formidable high-tech problem is not innovation—the problem is marketing new ideas.* In response to this, Carnegie-Mellon has merged elements of its graduate schools of engineering and business into a two-year curriculum leading to a combined degree in engineering and management.[14]

High Tech—*The* Agenda for America?

High technology has been portrayed and extolled as the road to success for the United States. But its benefits have too frequently been embellished. High tech is often billed as a powerful economic medicine for what ails the depressed regions of the nation's industrial heartland in the Midwest and as a stimulus to regions that have never fared as well industrially and economically. By some accounts, high tech is a mysterious potion that will do wonders if the recipient only drinks enough of it.

Consequently, it seems as if every state, every county, every city wants a piece of this high-tech action that will somehow propel or return every nook and cranny in the United States to industrial vigor. As the thinking goes, high technology will bring plenty of glamorous white-collar jobs and none of the pollution that goes hand-in-hand with the old basic smokestack industries.

Politicians and community leaders all over the United States pledge to get this thing called high technology. And many try hard to deliver on their pledges—competition for high-tech companies is fast and furious. Tennessee's pitch to high-tech firms is typical. It asserts boldly, "When it comes to high technology, Tennessee is getting down to brass tacks and giving you the moon and more." (Apparently, a great deal of U.S. space research was carried out in Tennessee, hence the moon metaphor.)

High technology is a great boost but plainly no panacea for what troubles the United States industrially. Yet, the levels of expectation surrounding high tech are raised too often by hype. Unfortunately, much of the public has been led to believe that high technology can achieve far more than it really can in fact deliver. Realistically, just what does high technology mean for the United States? Especially, what can it do in the way of creating new jobs and replacing those it is obsoleting? How many blue-collar workers can be trained to work in high-tech industries? Can all regions of the United States expect to participate in the high-tech boom?

High Tech without the Puffery

Realistic answers to these questions lead to the inevitable conclusion that high technology is one very essential agenda for America. Technological progress as a nation is immensely important for reasons of trade, national defense, health, and freeing citizens from mundane tasks like working on a noisy production line or doing other laborious work that is better handled by machines. But by no means should high technology be *the* agenda for the United States in the quest for economic growth and its beneficial byproducts. The vast majority of sites in the United States simply do not have the proper combination of elements to incubate or accommodate high-tech firms; nor are the prospects even marginally bright for most places getting what is necessary. What is more, if America as a nation depends on high technology to provide for reindustrialization and (net) new jobs, it will come up disappointingly short—most noticeably so in the next five to ten years.

Why do most areas of the United States have little or no chance to incubate or attract high-tech firms, much less to become high-tech centers like the well-publicized Silicon Valley in California or Route 128 in Massachusetts? A report by the National Governors' Association committee on technological innovation clearly illustrates why. It specifies four basic prerequisites that a locale must have if it is to spawn successful new high-tech endeavors. In addition to facilitating reasonably priced investment capital or seed money, the other three are:

A sufficient research base generating scientific and technical advances

A managerial structure with sufficient vision, experience, and know-how to transform good ideas into marketable products and services

A well-trained labor pool of scientists, engineers, technicians, and skilled workers

Significantly, the report pointed out that the hubs of high-tech innovativeness in the United States are proximate to the "finest research universities." The key word here is finest. It is not enough for a state or a city to have a university with an engineering school. Not just any engineering school will do, not even a good one. As J. Herbert Holloman, a professor of engineering at the Massachusetts Institute of Technology, has said, a world-class engineering school is needed to supply skilled workers.

At a meeting of the National Conference of State Legislatures, James Howell, a banking executive with expertise in high-tech location, identified California, Massachusetts, and perhaps eventually Texas as good parents to high-tech start-up ventures. A second group of states—Arizona, Minnesota, and North Carolina and its Research Triangle—have the ingredients to become good foster parents. "They won't start companies, but they will know how to take care of them when they get going," says Howell. According to Howell, "In the rest [of the states], the ball game's over. They should not even enter. They're not going to be a participant." Not many states act as if they buy his gloomy assessment of their chances in the high-tech sweepstakes. Consider some of these initiatives to foster and attract high-tech entrepreneurship:

Incubator facilities, such as Georgia Tech's on-campus Advanced Technology Development Center

Community partnerships among business, labor, government, and higher education, like Pennsylvania's Ben Franklin Partnership that helps establish technology centers

State venture capital funds to aid high-tech entrepreneurs

State training and retraining programs for producing technically skilled workers

Public and quasi-public organizations that assist high-tech entrepreneurs to get started by helping them evaluate ideas, develop prototypes, and conduct marketing research[15]

The majority of these efforts will not be fruitful. Most places in the United States do not have the remotest chance to become a viable high-tech area; they simply do not have the necessary ingredients. By definition, there are very few world-class engineering schools. In fact, the Great Lakes states have begun to look at themselves as a region in terms of providing a high-tech educational environment. Most states and large metropolitan areas will be able to showcase a few high-tech success stories here or there. But on balance, from a cost-benefit standpoint, the efforts will not have been worth it. The preponderence of state and local initiatives to promote high technology will be more productive if they are broadened to encompass entrepreneurship per se. High-tech jobs are but part of the solution to the erosion of basic manufacturing jobs in the United States. Low-tech and service ventures must contribute the most. The fact is, high-tech companies will concentrate in those relatively few places in America uniquely suited to their special needs. Other not-so-conducive locales are doing little more than chasing an elusive high-tech fantasy. One need only to review a recent issue of *Plants, Sites and Parks* to see how aggressively high-tech firms are pursued.

The "jobs" potential of high technology has also been oversold. Estimates by both the U.S. Bureau of Labor Statistics and Data Resources, Inc., indicate that the coming decade will see high-tech industries create less than one-half the jobs (730,000 to 1 million) lost in manufacturing (2 million) in the 1979–1982 period alone. Also, less than a third of the jobs created by high-tech firms will be technical. The majority of the occupations needed will be traditional ones, such as managers, clerical workers, operatives, and craft workers.[16]

High-tech industries create jobs, but they also take them away because of the productivity improvements their products make possible. A Congressional budget officer predicts that by 1990 there will be three million fewer manufacturing jobs in the United States. By the year 2000, robots are likely to be able to manufacture what seven million people do now.[17]

When it comes to retraining blue-collar workers from the basic industries to work in high-tech industries, there are major obstacles. There are a significant number of such workers who, once retrained, will still not be able to find employment. Already, there is an oversupply of technicians to operate robots.

Many of the blue-collar workers from the old smokestack industries are not readily retrainable. In the first place, the educational system in the United States is not up to the task. Budget crunches have taken their toll and generally have left educational institutions too weakened to gear up quickly for the retraining task. More limiting is the fact that the vast majority of the schools, colleges, and universities in the United States have neither the faculty expertise nor the facilities to adequately perform the needed retraining anywhere near adequately. A typical liberal-arts-oriented college, for example, assuredly cannot retrain displaced blue-collar workers for technical high-tech jobs, and many—perhaps most—universities are not too much better off. The universities in this country with engineering schools and *sufficient* computer staff and facilities are in a small minority.

A Red Herring

As important as it is, high technology has become something of a diversion—a red herring—from major problems confronting the United States. High technology has often been touted as an almost magical escape from the need for the nation to address the weaknesses that put the United States at a competitive disadvantage against foreign competition at home and abroad.

The basic industries in the United States cannot, or prudently should not, be neglected or abandoned in a wholesale rush to high technology. In the ensuing decade, there will not be a sufficient number of jobs generated by high-tech ventures to accommodate blue-collar workers displaced from the basic industries. There is also the consideration of national defense. Can the United States afford strategically to be reliant on other countries for the products manufactured by smokestack industries? Who is to say that today's trading partners might not be tomorrow's enemies or, more plausibly, be devastated by belligerent military actions? A survey of leading U.S. business executives conducted by Louis Harris and Associates found that three-quarters of them would be against further declines in America's basic industries if the declines would undermine U.S. defense. A sizeable majority of the executives polled favored tax incentives (e.g., tax-exempt debt) to assist basic industries to modernize and thus become more competitive.[18]

These are the kinds of important discussions and debates that all the high-tech publicity has tended to obscure. The high-tech dialogue must not allow government, business, and labor to avoid unpleasant questions like:

> What can be done about overregulation of business in the United States and the onerous red-tape impediments to innovation and productivity in *all* industries?
>
> Is it a coincidence that there are nearly as many lawyers in Washington, D.C. alone as there are in the entire country of Japan?
>
> What, if anything, can be done about the inordinately high wages and fringes of many blue-collar workers in the U.S. basic industries that cause a competitive price disadvantage for U.S. products, domestically and internationally?

What progress can be made in improving the real and perceived product quality of U.S. goods?

Which basic industries need to be saved for reasons of national defense, and how can it be done through a private-sector solution with the cooperation of government and labor?

Social Boon or Bane

With or without widespread modernization and labor-management cooperation in attacking the problems in the smokestack industries, the United States will continue to be transformed irreversibly to a predominately white-collar nation. Inevitably, the basic industries will be scaled down so that employment opportunities will reside mostly in the service and high-tech sections of society. This trend is hardly new; it has only been accelerated as of late. As long ago as the mid-1950s, the number of white-collar workers in the United States surpassed for the first time the number of blue-collar employees. But the trend has been given great momentum by recent technological breakthroughs in manufacturing automation (notably, robotics) and the loss of traditional labor-intensive factory jobs to foreign competition.

This latest version of an industrial revolution is causing unfortunate and often tragic social and economic maladies similar to those experienced in the original Industrial Revolution. But it is unavoidable. Places like the erstwhile steel center of the Mahoning Valley in Ohio can never return to the full-production glory days of yesteryear anymore than the displaced agricultural workers of the Industrial Revolution could or did return from factories to farms.

In the next decade, the social and economic costs will be considerable. As one writer puts it, "A generation of blue-collar workers may be stranded on their own shores."[19] Certainly that is true for the older blue-collar employees—those within 10 to 15 years of retirement. Government, business, and labor will cope with this problem, but cannot really solve it. The blue-collar dislocation problem—the structural unemployment—will linger until demographic and educational forces take hold. The reverberations from this latest industrial revolution, like its predecessor, will ultimately (10 to 15 years at most) atrophy as the baby-boom generation in the United States ages and increasingly gives way to the less numerous, more educated, and technologically adept baby-bust generation. Then, labor scarcity will actually be the problem confronting society.

What Is High Technology?

Definitions of high technology vary, ranging from the very subjective to the very quantitative. The U.S. Bureau of Labor Statistics classifies high technology according to the number of technical employees and the amount of research and development expenditures in a given industry compared to the average for all U.S. manufacturing. If an industry is to qualify as high technology, it must have twice the number of technical employees and double the R&D outlays as the U.S. average. Of the 977 industries assigned U.S. standard industrial codes (SIC), 36 qualify for the high-tech nomenclature (e.g., computers and computer programming, electronics, aircraft, drugs, and data processing). Another 56 industries are considered to be high-tech intensive, meaning that their R&D expenditures and technical employment figures exceed the national average (for example, some equipment makers in the electrical, printing, medical, and textile fields).[20]

The National Science Foundation (NSF) regularly reports research and development outlays applied in the manufacture of thirty product groups encompassing all U.S. manufactured goods. R&D expenditures are therefore used as a surrogate measure for high technology. Lester Davis, an official with the U.S. International Trade Administration, considers straightforward high-tech measurement techniques of this nature to be inadequate because they fail to account for indirect R&D inputs to manufactured products. In the field of chemicals, "The technology embodied in computers and instruments contributes immensely to the technology embodied in chemicals . . . the indirect R&D embodied in chemicals significantly understates their total technology intensity." Davis has suggested an elaborate input-output technique to overcome the problem he has identified.[21]

At the other extreme are subjective, judgmental definitions of high technology. Westinghouse Electric's general manager of advanced technology for its Defense and Electronic Systems Center, Eugene Strull, says that high tech is any technology that changes rapidly.[22]

The magazine *High Technology* published what is called a high technology index, which was developed by Bud Anderson, the editor and publisher of a monthly investment letter on high-tech stocks. The index consists of these products:

Analytical Instruments
Building Controls
Cable TV Equipment
CAD/CAM
Computer Memory Products
Computer Output
Computer Peripherals/Printers
Microwave Components
Microwave Equipment
Military Systems
Minicomputer/Distributed DP
Mobile Radios/Paging
OCR/Voice Recognition
Other Computer Peripherals

Computer Software/Services
Data Communications Equipment
Discrete Components
Electromechanical Components
Genetic Engineering
Home Computers
Integrated Circuits
Laser and Infrared Equipment
Mainframe Computers
Medical Equipment/Supplies
Medical Imaging Equipment
Other Medical Diagnostics
Pacemakers, Implants
Passive Components
Pharmaceuticals
Process/Industrial Controls
Security/Fire Systems
Semiconductor Manufacturing
Semiconductor Manufacturing Equipment
Telecommunications Equipment
Test Equipment
WP/Small Business Computers

None of these definitions is all-inclusive. Some products are consensus high-tech products; they are on everybody's list. Other products, all agree, are not high technology by any measure. A third group of products and services may or may not be defined as high technology, depending on who is asked or the measure used. For our purposes, definitional distinctions at the margin are unimportant and not worth debating. The book's contents are germane to management in any company that participates in a business with high-tech characteristics:

The business requires a strong scientific/technical basis;

New technology can obsolete existing technology rapidly; and

As new technologies come on stream their applications create or revolutionize markets and demands.

Why a Book on High-Tech Marketing?

A great deal has been made of the slippage in U.S. technological leadership. A number of contributing factors have been cited.

The National Commission on Excellence in Education has said that secondary education standards have been eroding in U.S. schools. As a result, American universities have been lowering their admission requirements. Since 1964, high school transcripts have decreased by 6 percent for chemistry and by 7 percent for algebra, while credits for driver education and remedial work in English have gone up by 75 percent and 39 percent, respectively. Less than a third of recent high school graduates enrolled in intermediate algebra. An American school year runs 180 days (of which the average student misses twenty days) in contrast to Japan's 220-day school year. Additionally, the Japan school day is longer. Not surprisingly, American College Test scores have been declining for high school seniors. In some subjects, there is a dearth of qualified instructors. Texas and Virginia have even offered bonuses to teachers of math and science, key subjects indeed when it comes to technological understanding.

Japan also produces five times the number of engineers as the United States. The National Science Foundation reports that, since 1965, the percentage increase in scientists and engineers engaged in research and development has increased by only 25.5 percent in the United States, compared to Japan (139 percent), Russia (140 percent), West Germany (100 percent), Britain (76 percent), and France (74.4 percent). In absolute numbers of scientists and engineers working in R&D, the United States remains a very distant second to the USSR. (It is widely reported that the United States has a shortage of engineers. It does not. What the United States has is a shortage of *young* engineers who are up-to-date in their knowledge. For various reasons, almost as many more experienced and older practicing engineers leave the occupation each year as enter it with new bachelor's degrees in engineering.)[23]

These are all serious problems in need of attention. Yet, there is trouble on the business side of innovation as well. Successful innovation requires far more than additional spending for science and math education, for training engineers, and for research and development. By comparison with the scientific/engineering side of innovation, there is truly a meager amount of available information and expertise about effectively commercializing the output of high-tech laboratories—in other words, about marketing high technology.

Just as there are problems with scientific and engineering education, business education is likewise suffering. For instance, over 20 percent of American business school faculty positions are vacant for lack of qualified instructors, vis-a-vis the more publicized 9 percent in engineering schools, and only about one out of every six business schools in U.S. institutions of higher education is accredited by the American Assembly of Collegiate Schools of Business. We believe that almost any high-tech firm's level of success correlates directly with its marketing acumen—its ability to conceptualize future needs and wants and to apply technology accordingly. R&D without direction from the business side is a "loose cannon" that may or may not be on target with the firm's business objectives and intended strategies.

A major task necessary to productive implementation of high-tech marketing is that of infusing and maintaining the Yankee ingenuity associated with miniscule entrepreneurial endeavors—the Carlson-Xerox endeavors of the world—in large, bureaucratic corporations. The overall philosophical tone necessary for high-tech marketing must originate with top management. The requisite entrepreneurial climate cannot exist and flourish without encouragement from the highest corporate levels. Like Mr. Morito of Sony, top management must be manifest advocates in words and deeds of the creative destruction of capital and the creation of markets and demands that may exist in an indeterminate future.

But that is not enough. Philosophy alone will not get the job done. There must be a commensurate high level of R&D spending relative to sales, as determined by historical and industry norms. But that is also not enough. As stated before, R&D efforts need to mesh with business objectives, which is precisely why marketing strategy is so vital in the high-tech firm. Alan M. Kantrow in the *Harvard Business Review* has referenced more than thirty books and articles that express the increasing perception by managers of the need to place corporate technological decisions in the context of overall business strategy. He says that technology has an inner logic that must be accounted for in a firm's strategic planning.[24]

Change is always the only constant, especially so in high-technology businesses. Marketing strategies require rapid adaptation as competitive environments oscillate from supply-side conditions to demand-side—as more and more competition enters an infant market and helps it mature—and back to supply-side when innovations obsolete existing technologies. A notable example of a problem faced by a high-technology firm that is an industry leader is the timing of when to outmode its own products through the introduction of new technology it has developed but held in reserve. (Rapid obsolescence by manufacturers' new technology is the reason that leasing is a popular practice in the television video cassette market.)

The role of high-tech marketing management is to apply technology strategically in the marketplace so that the firm gains a competitive advantage. This achievement requires close linkage between the high-tech firm's R&D and marketing functions. Richard N. Foster has conceptualized this linkage process as depicted in figure 1-1: The R&D Effort Portfolio. In his grid, the high-tech company's products can be classified into several "effort warranted" categories, according to the degree of technical/market fit:

Heavy emphasis—deserving full support, including basic R&D

Selective opportunistic emphasis—may be good or may be bad; requires a careful approach and top management attention

Limited defensive support—merits only minimum support[25]

As a rule, successful high-tech companies focus the efforts of their R&D people. R&D is not given a free rein to invent and develop a product,

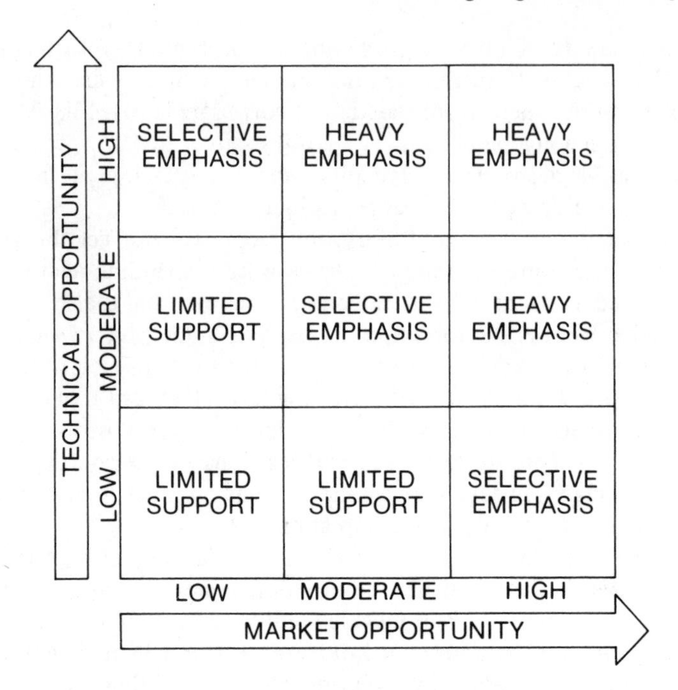

Source: Richard N. Foster, "Linkage Comes to United International." Reprinted from *Business Horizons,* December 1980, p. 70. Copyright 1980 by the Foundation for the School of Business of Indiana University. Reprinted by permission.

Figure 1-1. The R&D Effort Portfolio

service, or process for which the marketing side of the business will later attempt to find a profitable strategic market application. Quite the contrary. Possible market opportunities guide R&D's efforts. Like any company, the high-tech firm must ultimately answer to the decisions of the marketplace, where the consumer reigns supreme. Yet, there is an important distinction. The marketplace for high-tech products exists, if at all, in a future that is often beyond most people's present experiences. The thriving high-tech company is one that is perceptive and creative in response to problems or desires that are largely unknown to potential buyers or users themselves.

As contrasts in high-tech marketing, consider three firms—Texas Instruments, Timex, and Dravo Corporation. Texas Instruments, an archetypical high-tech, supply-side firm, succeeded for years on its technical know-how. However, lack of adequate technical/market linkage resulted in its failure to discern that supply-side market conditions in its consumer markets had changed to more mature demand-side environments with emphases on marketing expertise and savvy. One highly publicized ramification was the demise of its watch business (and now in-home personal computers) and a subsequent move to correct its shortcomings in marketing.[26] Whereas Texas Instruments' marketing prowess was not equal to its technological skills, Timex's dilemma was just the opposite. Timex, a prototype demand-side watch marketer, began encountering difficulties when it did not keep technological pace, specifically with the Swiss quartz models and with the Far Eastern manufacturers in digitals.[27] By contrast to these prominent examples of technological/market linkage gone awry, take the Dravo Corporation—a firm that has had the necessary coupling. Dravo attributes its competitive edge to the way it has developed technology and opportunely applied it in new ways in industry after industry. But the applications were not just in any old markets. Dravo carved out technology niches in domestic and international markets that its management evaluated as having high-growth potential.[28] Future market considerations drove the R&D, rather than the reverse.

Mr. Kantrow has commented in the *Harvard Business Review* that "The major unfinished business of the research literature is to provide managers with needed guidance in their formulation of a technology strategy . . ."[29] Conversations with executives in high-tech firms convinced us that Kantrow was right on target—there is a real urgency to know more about marketing high-tech products, from new product conceptualization through commercialization. That is why we so extensively researched the subject and then wrote this book on our findings.

What to Expect

It is not our intention to provide in-depth, technical discussions of specific marketing techniques, methods, and theories as they apply to high-tech companies. Detailed technical explanations are clearly beyond the purpose and scope that we have established for this book. The focus is heavily, although not exclusively, on what to do to make a company a competitive high-tech marketer, rather than on how to do it. The major topics covered in the forthcoming chapters are:

Staffing

Selection and placement of personnel in high-tech firms need to concentrate on seeking individuals who are entrepreneurial—innovative, creative, not averse to risk. Successful demand-side managers do not necessarily make successful high-tech managers, and vice-versa. What is a valued trait in one set of circumstances may be a negative in another. What are the traits shared by proven high-tech marketing executives? Are their backgrounds similar or diverse? Is there a pattern?

Organizing for Innovation

Segregated organizational arrangements, such as venture groups and project teams, which report directly to top management, may be critical in order to separate high-tech managers from the more cautious influence and control of demand-side managers. Segregating a high-technology unit might secure it the needed measure of exemption from the short-term performance pressures normally experienced by corporate profit centers. Hewlett-Packard, for example, decided years ago to establish a central research laboratory at corporate headquarters. Top management felt that R&D programs at the division level were too product-oriented and that not enough attention was being devoted to high-risk technologies which might have long-range fallout for the entire company.[30] Some sort of venture-team organization also encourages coordination and the checks and balances of a "devil's advocate" atmosphere, particularly between R&D and marketing. In what organizational ways can a high-tech firm improve the technological cooperation among its divisions and between its technical and marketing people?

Marketing Research

High-technology products, by their very nature, are of great risk compared to new products intended to fulfill existing market demand. Yet, marketing research techniques, if adapted and developed especially for evaluating the future market feasibilities of technological breakthroughs, can at least do a better job of providing improved linkage between R&D and the marketing/business side of a high-tech company. Take American Telephone and Telegraph. It has seen the need to go beyond conventional research to tailor what it calls the "Expanded New Product Research Process" for high-technology products. Its division manager for consumer market research believes that AT&T has thereby eliminated some of the needless risk in bringing high-tech products to market.[31]

There already exists a substantial amount of empirical evidence concerning diffusion of technological innovation. How is this body of research useful to marketers of high-tech products and services?

Marketing Decision Variables

The elements that the marketing manager has under his or her control and can manipulate are referred to in marketing parlance as the 4 Ps—product, price, place (distribution), and promotion (advertising, personal selling, sales promotion, and public relations). All of them are considered in upcoming chapters with careful attention given to choosing and discussing marketing concepts and techniques that are especially germane to high-tech companies. Collectively, when integrated into an overall marketing program for a company, these marketing decision variables are known as the company's marketing mix. Each firm's marketing mix is unique to that firm. As a cook tailors a recipe, so too does the marketing manager put the "controllables" together to suit individual tastes and situations (markets). Hence, the marketing mix of any two firms will be dissimilar as to the relative emphasis placed on the individual decision variables of product, price, distribution, and promotion. The marketing mix decision is extremely critical because it is instrumental in determining the degree of success experienced by a company in the marketplace vis-à-vis its competitors. Two firms with identical marketing expenditures in total dollar outlays may have very different results in the marketplace precisely because of the way each firm has developed its marketing mix. For instance, a firm's decision to allocate dollars heavily to advertising instead of upgrading product quality could be telling one way or the other; the decision may prove to be a success, but it also might be damaging. Since no company has an unlimited marketing budget, trade-offs between and among the marketing mix variables—the 4 Ps—are inevitable. It takes little skill to spend a lot of dollars on marketing; anyone can do that. The true test of the marketing executive is how those dollars are allocated among the marketing mix variables and, ultimately, what performance results in the marketplace.

Market Segmentation and Targeting

These concepts are closely intertwined with the marketing mix idea. Few companies, the large ones at any rate, have only one marketing mix. More normally, multiple marketing mixes are designed to fit the special needs and wants of separate groups of potential buyers—market segments. Usually, there are *n* marketing mixes for *n* segments. These groups (really subgroups of the total market) or market segments have been targeted by the company as potential customers. The aphorism that goes "different strokes for different folks" aptly describes the raison d'être of market segmentation and targeting.

Strategic Market Planning

Getting ready for tomorrow's contingencies is crucial for any organization or individual. But, by its very character, the high-tech firm operates in an environment that can change almost overnight with the discovery or application of new technology which obsoletes the old. Thus, can it really plan for any period but the near term? Yes, anyone who says it cannot confuses strategic planning with forecasting. Strategic planning is even more imperative in high-tech firms than in companies doing business in more predictable mature markets. Why? Because conditions are evolving so rapidly—conditions that can literally make or break a company quickly. The high-tech organization can be proactive. To a large extent, it can control its destiny; it need not be content to react to change.

Notes

1. David M. Smick, "What Reaganomics Is All About," *Wall Street Journal,* July 8, 1981, p. 20.
2. Adapted by permission. William L. Shanklin, "Supply-Side Marketing Can Restore 'Yankee Ingenuity'," *Research Management,* May-June 1983, pp. 20–25. Published and copyrighted by Industrial Research Institute, 100 Park Avenue, New York, New York 10017. All rights reserved.
3. Akio Morita, "Creativity in Modern Industry," *Omni,* March 1981, p. 6.
4. Susan Dentzer, "And Now, a Filmless Camera," *Newsweek,* September 7, 1981, p. 68.
5. "A Medical Marvel Goes to Market," *Business Week,* April 11, 1983, pp. 56–58.
6. Dentzer, p. 68.
7. See, for elaboration, William L. Shanklin, "Strategic Business Planning: Yesterday, Today, and Tomorrow," *Business Horizons,* October 22, 1979: pp. 7–14, and "Asset Redeployment: Everything Is for Sale Now," *Business Week,* August 24, 1981, pp. 68–74.
8. Derek F. Abell and John S. Hammond, *Strategic Market Planning* (Englewood Cliffs, N.J.: Prentice-Hall, Inc., 1979), p. 280.
9. Bill Abrams, "Booz Allen Looks at New Products' Role," *Wall Street Journal,* March 26, 1981, p. 23.
10. Smick, p. 20.
11. William C. Norris, "How to Expand R&D Cooperation," *Business Week,* April 11, 1983, p. 21.

12. Frederick Rose, "Many Firms Increase Basic Research Outlays after Years of Neglect," *Wall Street Journal,* September 3, 1981, pp. 1, 14.

13. Michael L. King, "Poll Suggests U.S. Industry Mismanaging Technology," *Wall Street Journal,* September 3, 1981, p. 21.

14. Thomas C. Hayes, "Brainpower: A New National Concern," *New York Times,* Section 12, March 27, 1983, p. 5.

15. Jack Magarrell, "Governors Warned about Weaknesses of Colleges in High-Technology Areas," *Chronicle of Higher Education,* March 9, 1983, pp. 1, 8.

16. "America Rushes to High Tech for Growth," *Business Week,* March 28, 1983, pp. 84–90.

17. Danforth W. Austin and J. Ernest Beazley, "Struggling Industries in Nation's Heartland Speed Up Automation," *Wall Street Journal,* April 4, 1983, pp. 1, 14.

18. "Executives Split on Saving Smokestack Industries," *Business Week,* April 18, 1983, p. 18.

19. David Halberstam, "The Quiet Revolution—Robots Enter Our Lives," *Parade Magazine,* April 10, 1983, pp. 16–19.

20. "America Rushes to High Tech for Growth," pp. 84–90.

21. Lester A. Davis, "New Definition of High-Tech Reveals that U.S. Competitiveness in this Area Has Been Declining," *Business America,* October 18, 1982, pp. 18–23.

22. "America Rushes to High Tech for Growth," pp. 84–90.

23. Andrew Pollack, "Engineers: A Dropout Problem for the U.S.," *New York Times,* Section 12, March 27, 1983, pp. 1, 12–15.

24. Alan M. Kantrow, "The Strategy-Technology Connection," *Harvard Business Review,* July–August, 1980, pp. 6–21.

25. Richard N. Foster, "Linking R&D to Strategy," *McKinsey Quarterly,* Winter 1981, pp. 35–52.

26. "When Marketing Failed at Texas Instruments," *Business Week,* June 22, 1981, pp. 91–94.

27. Jeffrey H. Birnbaum, "Failing Profit Prompts Timex to Shed Its Utilitarian Image," *Wall Street Journal,* September 7, 1981, p. 27.

28. Philip J. Berg, "Our Edge Is the Way We Apply Technology," *Wall Street Journal,* September 8, 1981, p. 3.

29. Kantrow, p. 18.

30. Robert T. Davis, Harper W. Boyd, Jr., and Frederick E. Webster, *Cases in Marketing Management,* 3d ed. (Homewood, Ill.: Richard D. Irwin, Inc., 1980), p. 228.

31. "Concept Trial Plus Marketing Research = Lower Risk in High-Tech Product Rollouts," *Marketing News,* September 18, 1981, p. 15.

29. Kantrow, p. 18.

30. Robert T. Davis, Harper W. Boyd, Jr., and Frederick E. Webster, *Cases in Marketing Management,* 3d ed. (Homewood, Ill.: Richard D. Irwin, Inc., 1980), p. 228.

31. "Concept Trial Plus Marketing Research = Lower Risk in High-Tech Product Rollouts," *Marketing News,* September 18, 1981, p. 15.

2 Marketing Organization and Sales Management

Structure is crucial, but unchanging structure is a snare and a delusion.

High-technology firms vary widely in terms of size, complexity, degree of specialization, and market composition. It is perhaps axiomatic that their most common feature—outside their emphasis on R&D—is their diversity. In semiconductors alone, the extremes range from the computer giants, IBM and Texas Instruments, to the smaller, highly specialized company, such as Insystems, Inc., a California company that plans to print integrated circuits with holographs.[1]

Thus, from the outset, it is evident that it is difficult to generalize about the marketing organizational forms that might be employed by a high-tech company. Although textbooks typically suggest four bases of marketing organization (product, customer type, geographic area, and function),[2] many factors regarding the company and its industry will need to be explored before we can consider the marketing component separately.

Types of High-Tech Firms

Realistically, a classification of the organizational types of high-tech firms would include:

The high-technology multinational company (MNC). Operating worldwide, these major companies (*Fortune* 1000) would produce predominantly or all high-tech products—consumer, industrial, or both. Illustrating this class would be IBM, Nixdorf Computer, Wang Laboratories, Fairchild Industries, and Eaton Corporation.

The consumer or industrial MNC with one or more high-tech subsidiaries. While predominantly mass marketing consumer or industrial products (non-high-technology), these *Fortune* 1000 firms have one or

---

The epigraph is a remark from Thomas J. Peters's article "Beyond the Matrix Organization," *McKinsey Quarterly,* Autumn 1979, p. 27. (Reprinted from *Business Horizons,* October 1979, p. 27.)

more subsidiaries involved in high-tech marketing. This class is illustrated by Champion Spark Plug Company (DeVilbiss) and Goodyear Tire and Rubber (Goodyear Aerospace).

The medium and smaller high-tech company. Primarily engaged in high-technology production efforts, these somewhat smaller companies are found in virtually every high-tech industry. Examples include Prab Robots, Inc. (robotics), Advanced Micro Devices, Inc. (semiconductors), and Hybritec (biotechnology).

The research laboratory company. While often involved in applications, a number of high-tech companies would be classified as predominantly research centers. Illustrative are the biotechnology companies Biogen S.A. and Applied DNA Systems, Inc.

The military applications-based high-technology company. Although often using military-related research spin-offs for other applications, these companies or subsidiaries of MNCs depend heavily on federal funding and sales. Examples are Hazeltine Corporation (information electronics) and CACI (software).

No classification system, of course, totally reflects the dimensions of the term *high-technology companies,* as there are many firms that would not easily fit into any of these categories. Furthermore, the medium and smaller high-tech classification is still a bit broad; it incorporates firms with $1 million or less in sales and some with total sales in the multidigit millions. The classification system does not really accommodate some companies, such as Corning Glass Works, which are involved in multiple high-tech industries. But still, it provides us with a clear indication of the difficulty in offering generalizations about high-technology companies.

Industry Development

Next, how important is the stage of development of the industry to our consideration of organizational structure and organizational problems? In other words, must an organization be structured differently or must it establish different organizational priorities if its industry is relatively new?

Clearly, there are various identifiable stages of development in any industry or form of industry evolution. Much attention is given to the s-curve of new product development in chapter 5, and this process often parallels the evolution of the industry itself. For our purposes, however, let us consider just three stages of development—stages that, for convenience, we simply refer to as: (1) the patent-driven stage; (2) the supply-driven stage; and (3) the demand-driven stage. As each new stage evolves in the industry, the firm must either adapt or drop out.

The Patent-Driven Stage. At the outset of a new breakthrough, firms tend to spend considerable effort vying for a degree of protected status. Today, breakthroughs rarely occur without prior fanfare and others being engaged in similar R&D. Their attention is focused on R&D, which is typically reflected in the company's organizational structure. Currently, this is illustrated by the biotechnology industry; an executive at Genentech, which has several hundred patents, indicates that patents serve as an "umbrella" that allows the smaller firm to develop.[3]

The Supply-Driven Stage. In the supply-driven introductory phase, the emphasis in the firm continues to come from R&D, as the new improvements and applications tend to push themselves on the market. The industry itself has not really been totally sorted out, as the number of producers continues to grow. In this immature industry setting, the emphasis is on R&D and technically skilled salespeople who must demonstrate how the product or process can be applied. The robotics industry is illustrative of this.

The Demand-Driven Stage. At this stage, the buyer has begun to have a better understanding of the product or process and can articulate his/her needs. The industry itself has become highly competitive and may be sorted out to the extent shown in the Porter Curve (p. 170) and its discussion. The demand-side has now moved into control, with the producers adjusting to the customers' needs rather than being able to push the product onto the market. Current illustrations of this stage are found in the computer chip, microcomputer, and software industries.

The marketing requirements and, concomitantly, the marketing organizational demands, are different in each stage. At this point, it is important to note that, at least in these three stages of an industry's evolution, a different answer is appropriate for the question "Organize for what?"

Organize for What?

We have stated that high-tech companies can be classified based on their size and complexity, and that the industries in which they operate can be ordered in terms of their stage of development. Each has important implications for the marketing organizational needs of the high-tech company. One final point that has similar importance for our discussion of such firms' marketing organizational requirements is that companies vary widely in their corporate objectives, and their objectives impact directly on the corporation's overall organizational structure, marketing goals, and marketing organizational decisions. All of the latter should be in concert with the overall purpose for which the company was organized and its specific long- and short-run objectives.

In their best-seller, *In Search of Excellence,* Thomas J. Peters and Robert H. Waterman, Jr., raise perhaps the most fundamental organizational design question of all: "Organize for what?"[4] This question is especially appropriate when considering corporate purposes and organizational arrangements for high-technology firms. Peters and Waterman see the most typical answer to the question from the large corporations in their research as "to build some sort of major new corporate capability—that is, to become more innovative, to be better marketers, to permanently improve labor relations, or to build some other skills which that corporation did not then possess."[5] These purposes have important implications for corporate flexibility and adaptability.

It has been hypothesized that larger corporations really lack the flexibility and creativity to produce breakthroughs. Rather, they may wait to determine the value of a breakthrough and acquire it, or simply fail to keep pace with the environmental changes that are occurring. It has been suggested also that it is the smaller firms that are truly inventive. Of course, the larger corporation does not necessarily have to accept this lack of inventiveness or adaptability to change; Peters and Waterman suggest examples of excellent companies that are "learning companies" and note that "the top companies have developed a whole host of devices and management routines to stave off calcification."[6] An interesting example of the magnitude of the flexibility and adaptability problem high-tech corporations face is offered by the Schering-Plough Corporation. In its 1982 Annual Report, the necessary multiple raison d'etre of the organization was expressed:

> Underscoring its fundamental mission of discovery aimed at developing novel therapeutic entities, the Company is moving determinedly to establish a leadership position in the emerging field of genetic bioengineering. At the same time, Schering-Plough continues to broaden its preeminent status in the cosmetics and personal care business through innovative product development designed to create new market niches and capitalize on the Company's quality brand name franchises.[7]

This major corporation cannot operate single-mindedly toward either existing products or R&D. Its organizational purpose must respond to both today and tomorrow if it is to remain viable. It must have an organizational structure that can successfully keep the company's multiple goals alive. In contrast, a company such as Hybritech—founded solely in response to a new breakthrough and industry—has been described as having the single purpose of "translating monoclonal antibody technology into commercial products."[8]

Now, what if we try to answer the question "Organize for what?" as might be appropriate for each stage of industry evolution? If we ignore the large corporation issue just described, our answers can suggest some rather specific purposes and objectives. In the patent-driven stage, the answer to the question is simply to establish the corporation's presence in the industry. For most new industries, being without a patent (patent pending) or access to licensing means that the company will not be in the industry at all. One would expect that many firms or subsidiaries of large firms will be totally R&D oriented at this stage and, therefore, R&D would dominate the organizational structure.

In the supply-driven stage, the shift is to applications and to product modifications that will maintain its state-of-the-art. The company's purpose will be to demonstrate its ability in the field. R&D will still receive significant attention in the organization, but it will now be coupled with marketing applications; and considerable attention will be focused on determining which applications will be most productive.

Lastly, in the demand-driven stage, the emphasis turns more to marketing and the answer to the "Organize for what?" question may be couched in terms of position in the industry and be something like obtaining the leadership role or dominant market share. At this point, R&D may receive minimal attention in the organization—perhaps not being directly represented in the top management hierarchy. The key skills may be related to production, financing, and, especially, marketing.

The point we wish to make is that the organizational design must permit the company to respond to the evolution of the industry. For the very large company, this is an ongoing process—especially when it is involved with multiple products and widely different industries. The excellent or at least successful firm must be able to adjust to the different early turns within one industry while operating at a mature level in another. Many large corporations, of course, find this extremely difficult, especially the earliest turns, and may prefer to wait to enter the industry until the actors become clear (supply-driven stage). At that point, establishing an agreement with one or more of the successful patent-race or early R&D winners—or acquiring one—may be a good strategy. Therefore, Square D Company acquired United States Robots, Inc., in the robotics field, while Schering-Plough developed an agreement with Biogen S.A. and acquired DNAX Research Institute to add to its already formidable biotechnology base.

Implications for Marketing

All of the discussion to this point has indicated the types of overall organizational arrangement problems that concern the high-technology marketer. Given this background, we have a better understanding of our constraints as we consider the marketing organizational structure.

Both the classification of the firm and the level of development of its industry have important implications for the high-tech company. For the existing company, especially the larger corporation, the formal structure will undoubtedly be in place and the opportunities to alter the established parameters may be limited. On the other hand, the newer company needs to consider the current stage of the industry in its organizational design, as well as its own classification.

In the subsequent section, marketing organizational problems (those that seem inherent to the high-tech company) and possible arrangements are explored. Supplementing the discussion are results of our research among select high-technology firms or industries. A major section of the chapter focuses on the sales force of the high-tech firm and the various needs for this especially significant portion of the marketing organization.

Marketing Organization

When evaluating the marketing organizational arrangements in any organization, there are a number of rather fundamental avenues that need to be explored. For example, what is the reporting level of marketing within the organization and is it appropriate to achieve the company's objectives? In addition, there are several organizational matters that should be assessed that are specific to high technology. The latter group can perhaps best be illustrated by the special need to interface R&D and marketing in the high-tech firm and the conflicts that such an interface can produce.

We consider several of these management or organizational issues in this section and relate their importance to the development of a successful marketing group within the high-technology firm. More specifically, we will address four concerns: (1) the reporting level of marketing; (2) the interface between R&D and marketing; (3) the relationship between marketing and sales; and (4) the locus of marketing decision making. While each of the four is important, the most significant is the reporting level issue. When the appropriate reporting level is established and the top marketing position filled by the right person, many of the other issues can be readily resolved.

Traditional Factors Influencing Reporting Levels

As we have said, there are a variety of traditional organizational designs that companies have employed as they structurally resolve their many external and internal relationships. To meet their external demands, the company has to be cognizant of the various outside forces that are of concern to them. These have typically included their customers or potential customers; their suppliers; their distributors, wholesalers, or retailers; their shareholders; and the general public. In addition, there have, on occasion, been special outside forces, such as governmental bodies or agencies, that have to be recognized because of their impact on particular industries. For some—such as those in the pharmaceutical or health care fields—it may have been the Food and Drug Administration and for others—such as companies dealing with lethal chemicals—the Environmental Protection Agency. The Securities and Exchange Commission and the Internal Revenue Service must also be included in this list.

Internally, the company has also needed to determine the appropriate relationships between the functional areas (i.e., finance and accounting, production, R&D, and marketing), so that it may achieve its objectives. In a sense, companies tend to establish a formal or informal pecking order for the functions, even though they often may appear to be equal. In fact, perhaps nothing shapes the overall direction of the organization more than this pecking order, which can determine if purchasing new equipment comes before an advertising campaign or whether the finance or legal department has the final say.

Not all organizational arrangements result from the careful planning implied by much of the management literature. Organizational charts are dotted with acquisitions that have never been fully integrated into the parent company or divisions that have hierarchical structures different from the rest of the company's units. Happenstance, short-term convenience, or a desire for separating problem areas do play roles in the internal relationships of some firms. But, in the main, the thrust of organizational design today seems to rather closely pattern traditional lines; for example, the medium-to-large company will use the functional division approach, the geographic division approach, the product line approach, etc., and perhaps overlay it with a matrix-reporting schema. Thus, as a result of these considerations and decisions, including a weighing of the plusses and minuses of each approach, one has a series of organizational arrangements that are designed to answer all authority and responsibility questions. In most basic marketing texts, the organizational charts are depicted with marketing operating at a comparable reporting level to finance, production, and two or three other key corporate functions. Each executive, including marketing, reports directly to the president or other chief operating official and is primarily responsible for all the functional units in his or her area of specialization. This arrangement holds regardless of whether we are considering consumer or industrial products.

Factors Influencing High-Technology Reporting Level

However, in some high-technology organizations, marketing may not be directly represented among the key company hierarchy; that is, the head of the marketing area may not sit on the company's executive committee or report directly to the president or chief operating official. Why might the chief marketing official be excluded from the top reporting level in the high-tech organization? There are several reasons that might be offered, and they typically relate to the fact that marketing plays, or is perceived as playing, a lesser role in such organizations. The first reason may be that the company is in an industry that is still in the patent-driven stage of development. Even though marketing can play a critical role in assessing markets and helping the firm position itself, the heavy R&D orientation of the company may cause it to ignore the value of marketing.

Second, there are undoubtedly heavily research-oriented executives, who tend to see marketing as merely advertising and promotion, and place it in a minor staff role in the organization. This kind of view has been cited at various times over the years as a distinct problem area in high-tech firms. To many scientists, engineers, and technicians, marketing represents a gray area of decision making, and they are accustomed to black-and-white decision making. Marketing becomes to them a necessary evil, and it is downplayed in the organization. We also consider this a potential problem area for the high-technology firm and discuss it later.

Third, there are companies that may actually have only limited need for marketing inputs due to the nature of their customers. Two examples are: the military products manufacturer operating primarily under U.S. government contracts and the research laboratory working under agreements with one or more major corporations. The former requires certain marketing-related expertise in lobbying and contract bidding, for example, but may have limited use for the types of marketing skills discussed in this book unless it sells to foreign governments. The research laboratory doing basic investigation and working under agreements with other producers may not be involved in determining potential markets or identifying marketable applications.

What Do the Study Findings Indicate?

In our studies conducted among proven high-technology companies, we did not find much evidence of a limited role for marketing. Much to the contrary, we found that marketing reported directly to top management in 82 percent of the firms (as shown in table 2-1). Only the finance function (with 92 percent) was more frequently found reporting to the chief operating officer of the firm. As might be expected from high-tech firms, R&D reported directly in approximately two-thirds of the firms. However, before dismissing the notion we raised earlier that the marketing function may be downplayed at the top, several other points should be mentioned. First, our research was conducted among the more successful and generally larger firms—companies that were beyond the patent-driven stage—and still, marketing did not report directly in roughly one-fifth of the companies. Second, we found that 90 percent of the senior marketing executives in our study had engineering or science undergraduate training and 25 percent had advanced degrees in engineering or science. Therefore, the pure marketing credibility or "soft area" issue would not be as apparent in these firms.

Where Should Marketing Report?

It is our view that marketing should be directly represented in the top echelons of high-tech companies; the experience of successful firms in virtually

Table 2–1
Frequency in Which Various Organizational Functions Report Direct to Chief Operating Officer
(percentages)

Function	*Percentage*
Finance	92
Marketing	82
R&D	67
Production	61
Personnel	45
International	39
Legal	37
Accounting	27
Public Relations	22
Other	14

every consumer and industrial market supports this position. Further, our study results suggest that a large share of high-technology organizations do include the senior marketing executive among their key decision-makers. (It is the 20 percent of the situations where marketing is not included that concerns us.) While the role of marketing is perhaps not being down-played as often today as it may have been in the past, our experience suggests that this can still be a problem, and one that the firm must guard against. Without the appropriate reporting level, it is quite easy for marketing to simply receive residual attention (i.e., a lower position in the pecking order that results in marketing needs receiving attention only after most other operations demands are met). For example, advertising expenditures might be the first line item cut if overall operating costs rise, even though an advertising campaign might be clearly justified if marketing were directly represented on the corporate executive committee.

Need for a Marketing and R&D Interface

A major criticism of high-tech firms has been their failure to develop a workable interface between R&D and marketing.[9] Additionally, when formal interaction has been established, conflict situations have often resulted.

Why is a marketing-R&D interface so essential? Basically, both groups play critical roles in the product development process, especially in high-tech firms. First, it must be recognized that the new product or process tends to be extremely complex and often has a number of potential uses. Some of these applications may be identified by the R&D group during their research, while others may be determined by marketing during the process of positioning the new product or process in an appropriate industry or niche within an industry. Each group has special expertise to contribute in this introductory stage of the product's development, and one or the other's absence from such deliberation can produce redundancy of effort, as well as numerous false starts. Since time is frequently a factor in the marketing of high technology, any unnecessary delays should be avoided.

Second, it needs to be remembered that high-tech products, unlike other industrial or consumer goods, tend to be supply-generated. Marketers often use the term *marketing concept* to refer to the development or modification of products based on research of existing needs. This tack has been described by some critics as producing new toothpaste colors rather than producing a true cavity-free toothpaste, as Proctor & Gamble did with Crest in the 1950s. Critics would say that the marketing concept or purely demand-oriented product developments result in product nuances, not product breakthroughs. Since the high-tech product or process breakthroughs represent truly new developments, R&D tends to dominate the firm in its early stage of development. Often, however, the R&D group lacks marketing skills and when such expertise is added, the R&D people cannot simply return to the laboratory assuming that their job is done. Their continued input in terms of application is essential.

Thus, we really see the need to involve R&D and marketing in product development or application efforts. However, conflicts may arise because each group often fails to appreciate the other group's expertise. We spoke before of the scientist seeing marketing as a "soft area"—one that lacks the scientific rigor of physics or engineering. Similarly, we must accept the fact that marketers often criticize the scientist as being too inflexible and lacking creative judgment. Frequently, as Joseph Jacomet and Joseph Wray have suggested,

> both groups experienced problems generally characterized by one of two extremes: either a lack of communications, appreciation, and trust or too much familiarity and friendship . . . [they indicate the groups] . . . prefer to go their own way which is certainly not the best way to reach organizational goals and objectives.[10]

Also, Jacomet and Wray add a slightly different perspective. They say that because marketing tends to hold a higher position in the organization, it may ignore R&D's possible role in the marketing development process. In this situation, R&D is viewed as a tool to implement marketing ideas rather than a point of organization of new product concepts. From either perspective, it is clear that the potential for conflict is present.

Development of a Marketing-R&D Interface

Given the desirability of an effective marketing-R&D interface, just how might it be achieved? There have been various organizational arrangements suggested for ensuring continued interaction between R&D and marketing. An approach to new product development followed by many consumer and industrial goods producers is the use of a venture team. This group typically includes representatives from marketing, production, R&D, etc. The team is involved in the product development task from idea generation to marketing introduction.[11] Certainly, this kind of arrangement, with some modification, creates the desired interface between R&D and marketing. The use of a formal structure, say a venture team, appears more satisfactory than a simple dotted-line relationship between the two; it requires the selection of representatives from both groups and a full-time commitment to the effort. The addition of other interested parties, finance and production, further demonstrates the importance of the team concept and helps to ensure that each will actively participate in the process.

Another alternative might be to establish a project task force. In a recent article, George Miaoulis and Peter LaPlaca have suggested the use of the project task force for product development in industrial high-tech companies when the technology is similar to a firm's existing technology and/or when the principal applications are in markets already served by a company.[12] It differs from the venture team in that the task force members have other responsibilities and, therefore, do not work full-time on product development. Again, the emphasis is on effectively integrating the strengths of R&D and marketing in positioning the product in terms of industry and application.

It is not necessary to employ a project task force or venture team approach to ensure an interface. But, it is necessary to provide some type of formal structure to ensure joint, complementary efforts of R&D and marketing, especially in the larger firm. The results of our studies of high-technology firms support this need to formalize the relationship.

As shown in table 2–2, item 1, for example, roughly half (48 percent) of the executives indicated that R&D has little or no involvement in planning the marketing strategy for new products. (In contrast, roughly three-fourths disagreed with the statement that marketing has little or no involvement in the activities of R&D.) Further, the view that R&D's role is presently limited in the marketing of high-technology products is supported by the other study results, as well. Apparently, therefore, an informal approach to developing the needed interface is likely to prove inadequate, and it is our recommendation that more formal steps be taken.

Table 2–2
High-Tech Marketing Executives' Views on Key R&D/Marketing Issues
(percentages)

Statements	*Agree*	*Disagree*
1. Typically, the R&D manager has little or no involvement in planning the marketing strategy for new products.	48	52
2. There is limited interface between R&D and marketing in high-technology firms.	23	77
3. Product planning is essentially an R&D function.	15	85
4. The research and development people are actively involved in marketing the new products they develop.	47	53

Selection of Marketing Organizational Structure

What factors affect the marketing organizational structure of the high-technology firm? Like any business organization, the structure of the marketing effort in the high-tech firm depends on both external (i.e., external to marketing) and internal factors. A primary consideration, of course, is the structure of the company. If the company has several divisions of subsidiaries and each represents a given product line or category of products, then it may be necessary to have separate marketing organizations for each. But a company may not have a variety of products or may prefer not to be structured along product lines. Instead it may choose to have the key functional areas of the firm (production, finance, marketing, etc.) all report to the CEO or president and have these officers assume the responsibility for their function throughout the entire organization. Under this type of organizational approach (functional), the chief marketing executive (vice-president or director) would handle all the marketing efforts of the company. For example, Micom Systems, Inc., a microcomputer producer with roughly $83 million in sales for 1983, employs basically a functional approach. (The vignette on page 42 describes the marketing/sales structure of Micom Systems, Inc.) Of course, the corporate and the marketing organizational structures are directly affected by external conditions, for example, the company's customers' needs, the geographic coverage of the company, overseas sales, its size in relation to its competition, and conditions within its markets.

For the widely diversified high-tech firm, it is often simpler to organize into divisions or subsidiaries based on product or industry categories. Champion Spark Plug Company, for instance, has a subsidiary (DeVilbiss) that produces and/or sells a variety of high-technology products. DeVilbiss, in turn, is organized into divisions based on such high-tech product categories as robotics and health care. This approach enables each division to develop fully the type of specialization needed, while being able to draw on the DeVilbiss headquarters in Toledo for any special support facilities. Or take the case of United Technologies with its total sales of $13.6 billion in 1982. This corporation is organized at the top into four organizational sectors; power, building systems, electronics, and Sikorsky/Inmont. Under the building systems sector are such large subsidiaries as Otis Elevator and Carrier Air Conditioning. The management and marketing needs for an organization like United Technologies clearly differ greatly from a Champion or a Micom Systems, Inc.

Marketing Requirements. The high-technology firm usually must be concerned with a number of marketing-related activities. Among its marketing

Organizing for an R&D-Marketing Interface: The Micom Systems, Inc., Case

The organizational challenges that face a rapidly growing company can be overwhelming. In its four-year corporate history, Micom Systems, Inc., has moved from $5 million to $83 million. This type of 25 percent per year annual growth would be unheard of today in many of the industrial manufacturing (so-called smokestack) giants, but is illustrative of the vitality that can be shown in true breakthrough high-tech companies.

But, as suggested, unless the company is able to handle such growth organizationally, its apparent success can result in its downfall. As we have suggested in this chapter, one of the possible reasons for a high-tech company's downfall can be its failure properly to interface its marketing and R&D activities, especially during the time when the company's focus needs to evolve from R&D to marketing. A most useful answer to this question is provided for us by Mr. Steve Frankel, the Vice-President for Marketing and Development, Micom Systems, Inc.

> My organization, Marketing and Development, has the corporate responsibility for all marketing (product planning, product marketing, product management promotional input, product manuals) and development (all engineering and R&D) activities performed by the corporation. Sales, both direct and through manufacturing representatives and distributors, and customer service, are handled by the Vice-President of Sales. Within Marketing and Development, we have three assistant vice-presidents who manage, on a product-line basis, the marketing and development activities of their respective product areas. In our case, that is data concentrators, data switching systems, and modems. Within each of these product-line organizations, we have individual marketing and development teams, so that as opposed to the typical competitive posture of these two functions, at Micom marketing and development groups work together to define new directions and solutions to the needs of their specific product areas. Once agreement is made, however, be it priority or product direction oriented, the marketing activity proceeds with the classical marketing functions while the engineering team begins the engineering process.
>
> As the project proceeds, project status information is available to both groups, and in-house quality and functional testing, beta test site testing, and product user manual reviews are also joint activities. So, in summary, the marketing and development teams work together on the front and tail-end of projects, and work independently from project go-ahead to the product testing and evaluation phase. Of course, if the development project does not proceed according to plan (i.e., cost or schedule) or market conditions change, joint efforts are again undertaken to reevaluate our position.

Just how successful has this approach been at Micom Systems, Inc.? This organizational structure has been in place at the company since 1982 and is felt to be most successful. The first full year of operation under this structure has been called Micom's most aggressive with regard to new product introduction by Mr. Frankel. He says, "twelve new products, equal in output to any two prior years, have been released. . . ." Mr. Frankel feels that "much credit for this success must be given to our unique organizational structure, which in essence has created a team-oriented psychology where a competitive psychology previously existed."

While the specifics of any organizational design tend to vary by company, industry, and the other factors we have suggested, the information provided by Micom Systems, Inc., and Mr. Frankel offer some valuable insights for the high-tech company. Mr. Frankel's last statement especially bears re-reading and perhaps should be printed on placards and placed in many executive offices within the high-tech company. A team orientation, in effect, should be a marketing and R&D goal of the high-tech firm.

---

requirements (i.e., those marketing endeavors that must be handled internally or handled by outside agencies) are those involving advertising; sales; marketing research; exporting; sales promotion and trade fairs; marketing planning; pricing; and the selection of a liaison with reps, wholesalers, dealers, or other middlemen.

While many high-tech companies use outside agencies to assist them in their advertising, marketing research, and exporting activities, few use outsiders to handle or advise them on their pricing, sales, and other marketing requirements. For example, in our broad study of high-tech firms, we found that 85 percent of the companies engaged the services of advertising agencies, 36 percent used marketing research firms, and 10 percent employed export management firms. However, the latter figure is misleading in that many of the firms have only limited export activities or interest at present.

In looking more specifically at high-tech companies' advertising activities in a separate study, we found that not only do virtually all these firms use agencies to handle their advertising, but also most use the marketing research services of their agencies. This offers one explanation for the rather limited use of specific marketing research firms noted above.

Since few firms employ outside services for their other marketing requirements, it is necessary for those activities to be assigned to in-house personnel. Just because an advertising agency is employed by a company does not mean that the company has no in-house advertising department. Most firms either have a small advertising department or at least have someone to work as liaison with their advertising agency. Our study found that the advertising departments normally contained two to six people, although two larger high-tech corporations had over fifty people each in their advertising departments.

Other Personnel. Without doubt, the largest number of marketing personnel in high-technology companies is involved in personal selling or sales promotion efforts. Personal selling is so essential to ongoing marketing efforts that much of the remainder of this chapter is devoted to personal selling and sales management activities. Therefore, little elaboration is needed at this point. Sales promotion involves a wide range of sales support efforts, ranging from the preparation of brochures and catalogs to trade show materials and displays. Trade shows are of great importance to many high-technology firms. In our encompassing study of high-tech companies, we found that 80 percent of the high-tech executives considered trade shows to be "a primary source for learning about competitors' new product development." A slightly different perspective was offered by a marketing manager in our robotics survey. This robotics marketer said that his firm "had to participate in the key robotics shows or everyone would think they were no longer 'in business'." Stated more positively, trade shows provide an important place for identifying potential customers, following the state-of-the-art developments, and reinforcing current customer relationships. For this reason, trade shows are an essential sales promotion endeavor, along with the preparation of sales materials. While one or more in-house individuals are involved in sales promotion, a firm's advertising agency, as well as trade show specialists, may be employed to assist them.

Increasing use is being made of marketing research by high-technology companies. Our inquiries indicated that a rather large proportion of the companies use some marketing research in their new product development efforts. Other studies, however, directed specifically to industrial marketing, have suggested that there is much more limited use of marketing research in the industrial products area than in consumer goods marketing.[13] Today, medium-to-large high-tech firms in particular employ marketing researchers, and their role often includes forecasting, as well as new product testing, pricing, etc. For example, the chemical marketing researcher is said not only to be involved in investigating markets, but also in the analysis of factors affecting manufacturing economies. He or she gives attention to the latter "in order to provide better price forecasts and more accurate assessment of competitiveness," according to the head of the European Channel Marketing Research Association.[14] Other personnel will be needed in most high-tech companies to select and provide liaison with middlemen, whether retailers/dealers or distributors/reps. As we see in our sales management section, many companies include various service personnel and applications specialists in the marketing or sales group.

Marketing Executive Qualifications

On several occasions, we have made or will be making references to the fact that the marketing efforts of the high-tech company need to have certain flexibilities, as the firm and the industry move from the new product breakthrough stage to ultimate market maturity. At one point the firm's marketing activities may need to emphasize R&D, product (patent) development and/or applications, and have a marketing head versed in these areas. Later it simply may need the pure marketer, who may have little or no scientific background. A case in point was the recent appointment of James J. Morgan as chairman and chief executive of Atari, Inc., and the appointment of John Sculley to a similar position at Apple Computer. Morgan had previously been vice-president for marketing at Philip Morris, Inc., while Sculley was president of Pepsi-Cola. Interestingly, Morgan claimed to know little about computers at the time he accepted the position and indicated that he wanted to model Atari after Philip Morris.[15] Since the computer industry has been engaged in that transition to growth or demand-side, a switch in leadership orientation by these firms should not be surprising. It merely reflects the recognition that scientific or engineering competence is less important at this point in the industry's development than marketing savvy. Marketing has been called the critical factor for success in the personal computer field today.[16]

Conversely, of course, is the reservation that the marketing-oriented president or marketing V.P./manager may be too unsophisticated to recognize or encourage real product development. Or, as Roger C. Bennett and Robert G. Cooper have suggested, the marketing-oriented executive may be too preoccupied with the needs of consumers "expressed in the consumer's own terms"[17] and, of course, limited to the consumer's own frame of reference. Dr. Edward Ungar, the head of Battelle Memorial Institute's Columbus Laboratories, has been quoted as stating quite pragmatically that "Every nugget of an idea should be explored with the dynamics of the market in mind. And after identifying a need, invent to solve the problem. This has far greater impact than technical wizardry in the laboratory."[18] But, if this view were taken too literally, the concern that R&D "will simply be a tool to implement marketing ideas"[19] might arise and many important breakthroughs lost.

Given these counter-views regarding top management qualifications, just what type of educational background or professional experience did the executives in our studies have? And how did they see this issue? First, in response to the statement "In high technology industries, a firm's marketing manager should have a technical background. . . ," some 90 percent of the general survey respondents indicated their agreement.

As we indicated in our introduction, the executives who responded to our general survey came from several industries (biomedical, software/computer, etc.) and were either the president (27 percent) or the senior marketing executive (57 percent were marketing V.P., manager, etc.). Some four-fifths of them had an engineering or science undergraduate degree and one-fourth had advanced degrees in science or engineering. Another one-third, however, did have an MBA degree. Most appeared to have spent a good part of their careers in technical-related posts, and therefore, they tended to have the technical tools that some feel are needed by the high-tech executive. Whether they will be able to cope with the transition problem raised earlier remains to be seen.

In our study conducted among marketing executives in the robotics industry, 68 percent had engineering or science undergraduate degrees and the remainder had business or liberal arts backgrounds. Thus, the robotics field, which is still a relatively new industry, tends to have the technical-trained marketing manager. With the current forecast of an imminent industry shakeout, the question of whether technical backgrounds offer the degree of flexibility needed to handle the transition into demand-side marketing also remains to be answered. Will some in the industry, in fact, turn to the consumer field for leadership, as did Atari or Apple? This is not likely. Since the robotics field is industrial-related, it is doubtful that such a move would seem viable. However, future selection from among successful marketing executives in demand-driven industrial products fields could occur if some robotics companies began to see their profits erode, as did Atari in 1983.

Personal Selling and Sales Management

The most common ingredient shared by high-technology marketing organizations is their use of personal selling. Even the smaller firm tends to have its own sales force, which is often combined with the use of middlemen to provide coverage for its markets. Given the high-tech marketer's penchant for having its own sales force, top management needs to be concerned with the sales reporting line and with the whole range of sales management functions.

Marketing versus Selling

Various organizational relationships within the marketing area have been debated over the years. "Where should advertising or public relations or new product development report?" is illustrative of the type of questions asked about organizational reporting arrangements. No reporting-line question, however, is more widely debated than the question of whether sales should report to marketing or, alternatively, should be positioned independently in the organization. In the latter structure, sales would have its own direct representation in the top management hierarchy.

The two basic alternative reporting approaches for sales are presented in figures 2–1 and 2–2, although there are obviously multiple variations of these two alternatives. In figure 2–1, the sales manager is depicted as reporting to the marketing vice-president or top marketing official, as are the advertising, marketing research, and other functional marketing heads. In contrast, figure 2–2 depicts the alternative approach in which marketing and sales have an equal reporting level.

How significant is this marketing versus selling reporting-line issue? Certainly, there have been successful firms represented by both approaches. However, the arguments tend to favor the type of arrangement described in figure 2–1. As is discussed in detail in our chapter dealing with presentations (chapter 6), it is essential to coordinate tightly the company's advertising and personal selling efforts. This can be more readily effected if they are both made to report to someone concerned with the firm's overall marketing efforts. Similarly, there is occasionally the opportunity to make certain trade-offs between advertising, personal selling, and other marketing activities. For example, a decision may need to be made regarding

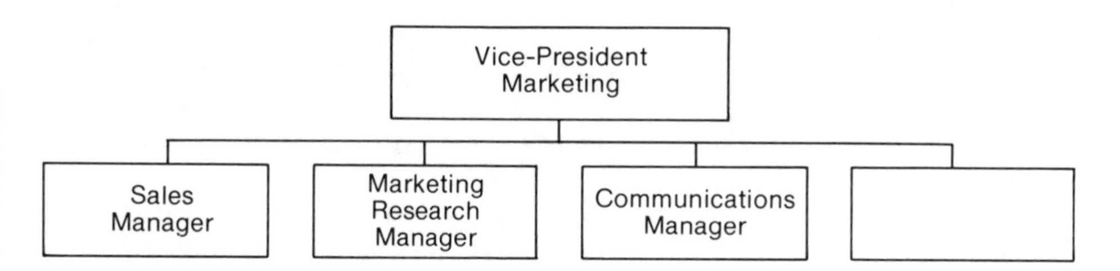

Figure 2–1. Marketing-Directed Approach

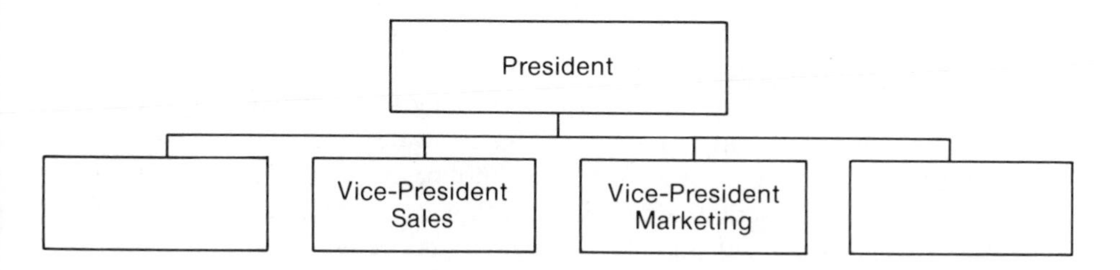

Figure 2–2. Marketing-Sales Equality

whether to increase the firm's advertising expenditures or to add to its personal selling staff. A marketing executive should be in the best position to answer this query dispassionately. It seems that the strongest argument that can be given for a separation of sales and marketing is that it emphasizes the importance of selling in the organization. Other arguments, often autonomy-related, merely seem to support the reasons why it has often been necessary to bring sales under marketing.

Our study of the robotics industry provides an interesting illustration of the reporting-line issue. In a little over one-third of the firms (37 percent), the overall responsibility for marketing the product fell to a V.P. for marketing, while a marketing manager was responsible in an additional 12 percent of the companies. In about one-fourth of the firms (23 percent), a national sales manager or sales manager had similar overall responsibilities. One often-used approach in 15 percent of the companies was the V.P. of Marketing and Sales title. In our more general high-technology survey, most carried a marketing-related rather than a sales-related title.

Do High-Tech Sales Differ from Other Types of Selling?

When one considers high-tech selling, two important considerations come to mind. First, there is the need for the salesperson to have technical training or direct access to technical expertise or assistance. Second, there is the need for the sales function to be closely related to and concerned with the service function in the organization. This need is illustrated by the Corning Medical subsidiary of Corning Glass Works. As shown by its organizational chart (figure 2-3), the company gives considerable attention to its service function. (For a complete discussion of Corning Medical, see the vignette in this section.)

The need for technical expertise (or access to expertise) on the part of the high-tech salesperson affects various sales management decision areas. For example, it influences the type of salesperson selected, the type of training he or she receives, the company's compensation plan, and even the size of territories assigned to the salesperson. (Each of these decision areas is explored in greater detail later in this chapter.) Furthermore, it may imply the need to link the application specialist with the salesperson, a practice widely employed in robotics sales.

The need for service per se varies according to the product or process and the category of customer—many large customers may provide for their own service needs. Two points that should not be overlooked are (1) that the company's service activities provide continued interaction and feedback from customers and (2) that the company's service activities may help to relieve the customers postpurchase (or cognitive) dissonance. Regarding dis-

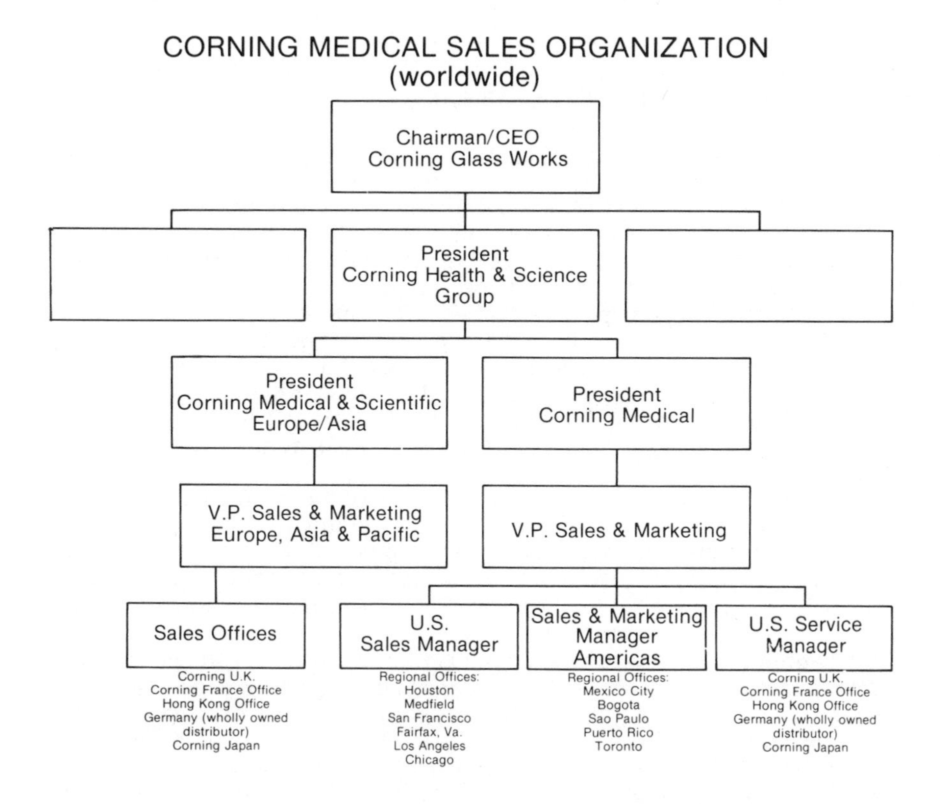

Figure 2–3. Corning Medical Sales Organization (worldwide)

sonance, research has shown that even the veteran industrial purchasing agent may have postpurchase doubts about whether or not he or she made the right decision after making a major purchase.[20] While no data are available specific to high-tech companies, the very nature of the rapid state-of-the-art developments in industries such as microcomputers, software, and robotics would suggest that they are ripe for postpurchase dissonance problems. Miland M. Lee, writing in *Business Marketing,* not only supports this view of the importance of service, but also indicates it to be a crucial element in the high-tech marketers' mix and a means to differentiate the company.[21]

Sales Training. There are several factors to consider when discussing the training needs of the high-tech firm. For example, some firms will be involved primarily in training new salespeople, while others are more involved in retraining (or offering continual training) to an existing sales

Corning Medical Sales Organization

Corning Medical, a major division of Corning Glass Works, employs its own sales force for the bulk of its domestic and much of its European and Japanese investment selling activities. (Further discussion, page 157.) The remainder of its sales operations in the United States and the rest of the world is handled by closely directed distributors or agents. The latter often handle a single product line and may represent Corning exclusively. As shown in figure 2-3, all selling activities in the United States and the Americas, including Canada, report to the Vice-President of Sales and Marketing, Medfield (Massachusetts), while the remainder of the company's worldwide operations are directed by a Corning Limited (United Kingdom) counterpart. Corning maintains direct linkage with its sales force/distributor or agent network via a series of regional (United States) or international sales offices.

In addition to their primary sales role, one of the principal functions of the distributors or agents is to maintain appropriate inventory levels, especially for the consumable products. To complement these efforts, Corning maintains an international as well as domestic customer service group.

Sales Force Characteristics. The typical Corning Medical salesperson has either a biological science and/or business administration academic background. The sales force's training is coordinated out of Medfield and includes both technical product training and in-house developed sales programs. As noted before, the sales force is organized, and therefore, trained, on a product-line basis. The principal responsibility of the sales force is selling, that is, determining and meeting customer needs. Feedback from the sales force to the regional office (and ultimately home office) regarding new product needs or new applications for current products is provided by weekly sales reports. Other suggestions are determined by the home office through monitoring field service reports which indicate potential problems. Also, product managers make frequent trips to hospitals, with salespeople, in search of new ideas.

International Operations. Basically, Corning Medical's international activities are handled from two centers, London and Medfield. Medfield's sales and marketing manager for the Americas is responsible for the sale of all the company's product lines for Puerto Rico, Canada, Mexico, and Central and South America. He directs and manages the sales offices in Toronto, Mexico City, Sao Paulo, and Puerto Rico. These offices are responsible for distributors or agents in their assigned territories. The bulk of the products sold in these areas is exported from the United States and often creates unique shipping and U.S. customs problems, as well as local import restrictions and currency difficulties. (All international financing arrangements are handled from CGW corporate headquarters in Corning, New York.)

Corning Limited, a wholly owned Corning Glass Works subsidiary, is the non-United States/Americas counterpart of Corning Medical. It handles the same product lines as Corning Medical and has a production plant in Sudbury (Great Britain) and has selling and service responsibilities for

Europe, Africa, the Middle East, Pacific Basin, etc. Corning Limited maintains sales offices which direct the company's own sales force and/or distributors or agents in their area of responsibility. For example, the Hong Kong office (and its area manager) is responsible for all the Pacific Basin (except Japan) and the Middle East. Like the Americas group, it operates through local distributors (or agents), who maintain inventory in their warehouses and handle all sales, training, and field service.

---

force. Similarly, the question of product or process-oriented training versus sales-oriented training needs to be resolved. While a company may feel that its salespeople, especially its new salespeople, need both product- or process-oriented *and* sales-oriented training, a shortage of time or funds may force them to opt for one or the other.

Then there is the question of which type (or types) of training to employ. Does the salesperson need to receive formal, classroom training, or is some form of on-the-job training, coupled perhaps with audio or video cassettes or other supplementary techniques, adequate to meet the company's needs? Naturally, the type of training indicated also depends on the company's hiring policies. If the company attempts to fill its sales force with individuals who have industry sales experience, its training requirements are likely to be far different than those firms who rely on recent college graduates as their source of new salespeople.

It was reported recently that Sperry Univac's strategy for marketing its new Sperrylink system placed very heavy reliance on its sales force. To overcome the difficulties of selling in a different market, Sperry found it necessary to retrain its current sales force and to hire talent from competitors. Over a relatively short time, the company planned to rotate its entire sales force through its training center.[22] By comparison, Monroe Systems for Business employed a video network to provide product and sales training for its United States and Canadian salespeople when it made a major product line expansion in September 1981.[23] These two companies illustrate the different approaches that may be followed depending on the company's perceived needs.

In our own research, we found that most of the firms in the robotics industries provide training programs for their respective sales forces. (Roughly four-fifths offer training.) The average length of these programs is two weeks, and the primary focus of the training tends to be product knowledge rather than selling techniques. Since many of the companies rely heavily on their application support department to assist in the sale, this focus on product knowledge indicates the importance placed on technology and technical expertise in this particular industry.

Compensation and Territory Assignment. Two important sales management decision areas are the sales force's compensation plan and the approach used to assign territories and accounts. As most sales management literature correctly suggests, there are normally no greater sources of sales force morale problems than compensation and territory inequities. Therefore, much of the sales manager's (and marketing executive's) time needs to be devoted to ensuring that the company's salespeople are positively motivated by the compensation and territory assignment approach that is employed.

There are numerous methods that are followed in territory assignment and they range from arbitrary assignments to what might be termed quantitative or empirically based techniques. An example of the latter is the use of operations research procedures to attempt to deal with the multiple variables that influence the quality of a territory and to resolve situations of inequity.[24]

Compensation plans all have considerable similarity. The basic question concerns whether or not the plan includes some means for directly rewarding performance (incentives) or instead, features security and control. The straight salary approach, unaccompanied by any bonus or commission, offers the salesperson security and the company the ability to control the individual salesperson's activities. Regarding the latter, a company can hardly expect the sales employee to spend a day completing reports if he or she is working on commission, but can direct such actions if the person is salaried. By basing the salesperson's compensation totally or in part on his or her sales performance, the company is providing incentive to perform at or above some established level. Obviously, there are trade-offs between the fundamental compensation approaches that the high-tech firm must consider. There is really no more critical sales force decision than the question of the compensation approach selection.

The robotics industry research again offers some interesting insight to this decision. We found that some 42 percent of the companies employed the straight salary approach, while the remainder tended to use salary plus bonus. Fewer than 10 percent used straight commission. These results are consistent with studies in other industries, which have indicated a strong trend away from the straight commission approach.

A third important decision area is sales prospecting; particularly, how sales prospecting relates to training. An extensive discussion of this topic is presented in the vignette entitled "High Tech Can't Forget Sales."

Sales Force Qualifications. Virtually all the literature pertaining to high-tech marketing has indicated the need for salespeople with engineering or scientific backgrounds. A thorough review of the employment advertisements, college recruiting announcements, or job descriptions, substantiates this qualification. A technical background requirement for salespeople has been especially noted in biotechnology, electronics, medical instrumentation, ceramics and robotics, to name a few. In our queries, we found that the typical educational background of the robotics salesperson included an

High Tech Can't Forget Sales

In our study on the robotics industry, we found that field sales personnel, current and potential customer interaction, and trade shows were felt to be most valuable in identifying potential markets by the robotics executives. However, we looked with considerably more depth into the question of high-tech sales prospecting in an article in the November 1981 issue of *Industrial Marketing.* This article, "High Tech Can't Forget Sales Prospecting," reported the findings of an in-depth study conducted among 75 sales representatives for a single major domestic computer company. (While this firm is selling primarily for technical and scientific applications, we feel confident these results apply to prospecting in most high-technology firms.) In particular, the company employs salespeople with engineering backgrounds (degrees), which we commonly find to be the practice in so many of the high-technology fields.

The results of the "source of names" and "methods of initial contact" phases of the research are presented in tables 2-3 and 2-4. However, to add insight to these findings, we specifically looked at the sources employed by the highest quota attainment salespeople (150 percent plus). We found that this elite group relies far more heavily on referrals, business-related articles in newspapers and trade journals, and civic meetings and cocktail parties than do the less productive groups. Perhaps most important, these highly productive salespeople have a higher regard for prospecting than do the remainder of the sales force.

Other key findings of the study include:

sales engineers with little experience rely more heavily on ineffective sources and contact methods

sales engineers with little experience initially choose contact methods that permit them to avoid face-to-face contact with prospects

salespeople with experience rely more heavily on face-to-face techniques, such as cold canvassing and asking for referrals, for initial contacts

salespeople in the highest experience category (seven-plus years in selling) spend a smaller portion of their time on prospecting

sales quota systems in which companies increase the quotas of high performers appear to be self-defeating (the highest experience category clusters around the 100 percent–120 percent performance level to avoid ever-increasing quotas)

salespeople with three to four years experience are the highest performing group

Based on our in-depth study of prospecting, we offer the following recommendations for improving sales force performance:

provide more prospecting assistance, such as having a sales manager (or more experienced salesperson) accompany the newer sales engineers on prospecting ventures, setting up customer educational seminars, and making specific business lists available for each territory

---

engineering degree. Some 77 percent of the robotics executives indicated that an engineering major was required for their sales force. This finding is certainly not surprising, since all our high-tech research has indicated that most individuals involved in high-tech research marketing have either an engineering or science undergraduate background. The only exception was the advertising and promotions area, and even there many did have technical back-

Table 2–3
Sources to Find Names of Prospects

	Percentage of Sales People Who:	
Source	*Often or Occasionally Use Source*	*Almost Always Find Source Effective*
Salespeople from your firm (those *not selling* in your line of business)	93%	48%
Referrals from satisfied customers	91	50
Salespeople from your firm (those *selling* in your line of business)	88	24
Engineering departments of potential customers	85	21
Obtaining one or more names from each person you contact during the day	63	25
Responses to your company's ads in trade publications	59	4
Local trade shows	57	8
Civic meetings, cocktail parties, and other social events	49	2
Purchasing departments of potential customers	48	3
Your company's files on firms located in your territory	48	12
Industrial directories	45	8
Business-related articles in newspapers or journals	31	1
Lists bought from list brokers	27	5
Local chambers of commerce	21	5
Professional society meetings	19	1
Discussion with noncompeting sales people from other firms	9	1

Table 2-4
Methods of Initially Contacting Potential Prospects

	Percentage of Sales People Who:	
Method	*Often or Occasionally Use Method*	*Almost Always Find Method Effective*
Call prospect on phone for appointment	99%	44%
Invite prospect by mail to seminar	92	22
Invite prospect by phone to seminar	87	24
Hold seminar at your company's office	81	28
Hold seminar at local hotel	76	23
Hold seminar at prospect's site	68	43
Invite prospect via newspaper ads to seminar	45	3
Mail letter of introduction about you and your company	44	1
Try to see prospect in person without appointment	41	1
Send locally written newsletter about your firm to prospect	37	8
Use radio ads to invite prospects to seminar	24	1
Hire personnel to conduct calling campaigns for appointments	7	1
Walk in to see prospect without appointment but with token gift	3	1

Source: Robert J. Pacenta, John K. Ryans, Jr., and William L. Shanklin, "High Tech Can't Forget Sales Prospecting," *Industrial Marketing,* November 1981, pp. 78, 80. (*Industrial Marketing* is currently titled *Business Marketing.*) Reprinted with permission.

grounds. As our discussions of new product development, distribution, and pricing in the following chapters will suggest, this focus on technical background rather than business or marketing (or a combination of the two) has certain strengths and weaknesses for the long-term viability of the high-technology company.

Notes

1. "Optics Fend Off New Methods of Making Chips," *Business Week,* November 22, 1983, p. 40B.
2. William Lazer and James D. Culley, *Marketing Management* (Boston: Houghton Mifflin Company, 1983), pp. 261–62.
3. Tamar Lewin, "The Patent Race in Gene-Splicing," *New York Times,* August 29, 1982, p. F-4.

4. Thomas J. Peters and Robert H. Waterman, Jr., *In Search of Excellence* (New York: Harper & Row Publishers, 1982), p. 8.

5. Peters and Waterman, pp. 8–9.

6. Peters and Waterman, p. 110.

7. *Schering-Plough 1982 Annual Report,* inside front cover.

8. Lawrence Prescott, "Hybritech: Portrait of a Monoclonal Specialist," *Bio/Technology,* April 1983, p. 157.

9. Joseph A. Jacomet and Joseph A. Wray, "Improving R&D's Interface with Management, Marketing and Manufacturing," *Adhesive Age,* August 1981, p. 29.

10. Jacomet and Wray, p. 29.

11. Richard T. Hise, Peter L. Gillett, and John K. Ryans, Jr., *Basic Marketing: Concepts and Decisions* (Cambridge, Mass.: Winthrop Publishers, Inc., 1979), p. 247.

12. George Miaoulis and Peter J. LaPlaca, "A Systems Approach for Developing High Technology Products," *Industrial Marketing Management,* October 1982, p. 260.

13. Glenn V. Ostle and John K. Ryans, Jr., "Techniques for Measuring Advertising Effectiveness," *Journal of Advertising Research,* June 1981, p. 22.

14. Sandra Heathcote, "Conservation," *Chemistry and Industry,* April 5, 1980, p. 252.

15. Leslie Wayne, "Philip Morris' Marlboro Man Switches Over to Pac-Man," *New York Times,* July 24, 1983, p. F-7.

16. "The Coming Shakeout in Personal Computers," *Business Week,* November 22, 1982, p. 75.

17. Robert C. Bennett and Robert G. Cooper, "The Misuse of Marketing," *Business Horizons,* November–December 1981, p. 54.

18. "Engineers Told: Invent to Solve Problems, Be Sensitive to Markets," *Elastomerics,* December 1981, p. 71.

19. Jacomet and Wray, p. 29.

20. Richard Hise and John Ryans, "Post-Purchase Dilemma—Did I Make the Right Decision?" *Midwest Purchasing,* December 1972, pp. 13–15.

21. Miland M. Lele, "Product Service: How to Protect Your Unguarded Battlefield," *Business Marketing,* June 1983, p. 69.

22. Philip Maher, "Sperry's Bold Gambit," *Industrial Marketing,* January 1983, p. 34.

23. Joseph W. Ardwody, "How Monroe Systems for Business Harnesses Video Sales Training Techniques," *Business Marketing,* June 1983, p. 90.

24. Robert C. Ferber, "Using Operations Research for Sales Territory Management," *Industrial Marketing,* November 1981, p. 66.

3 Positioning and the Selection of Target Markets

Multinational companies that concentrated on idiosyncratic consumer preferences have become befuddled and unable to take in the forest because of the trees.

Contrary to popular belief, successful marketing rarely involves much luck. Rather, it requires effective planning. As a saying goes, luck is infatuated with effort. Nowhere is planning more important than in the identification of the most appropriate and profitable market opportunities for a high-technology firm's products or processes.

No marketing activity is more significant than the delineation of the priority prospects for each product, process, or service of the corporation. And this axiom is true for both consumer and industrial high-tech marketing. A DuPont executive based in Geneva, Switzerland, recently mentioned that his most important marketing tool was his mailing list containing all the key chemical buyers in Europe. What he was saying was that his firm had identified the potential customers that composed its European target market, the potential buying group offering the primary opportunities for its chemical products. Therein lies the reason he considered the list to be so important.

Certainly, the activities associated with identifying the corporation's primary opportunities or target markets have received considerable attention by marketers. As markets for specific products or services become more and more mature, we find firms even searching for singular niches or subsets of the market where they feel they have competitive advantages and expertise. The term *market segmentation* was developed in the marketing literature to describe the activities associated with target market selection, that is, "breaking the total market into logical market segments that differ in their requirements, buying habits, or other critical characteristics."[1]

Consumer goods companies, especially, have become increasingly adept in isolating potential market groups or segments and then searching for these groups' specific needs and wants. For example, we have seen this

---

The epigraph is a remark by Theodore Levitt from his article "The Globalization of Markets," *Harvard Business Review,* May–June 1983, p. 92.

kind of demand-side research result in the soft drinks Like, Pepsi-Free, and others that are caffeine-free, sugar-free, or both sugar- and caffeine-free. Similar illustrations could be provided in most consumer goods fields and for some industrial products.

But unfortunately, little of the research that has been done in the area of market segmentation has direct relevance for the high-tech, supply-side-oriented producer. Before describing the specific target market identification problems of the high-tech marketers, we will first examine what market segmentation by the demand-side marketer entails. And in so doing, we will see what Theodore Levitt means when he says companies "customize products for particular market segments."[2]

The Art of Market Segmentation

Market segmentation activities designed to identify target markets or customer categories for consumer goods producers may range from simply classifying the market into various demographic categories to conducting extensive psychological- or sociological-type research. With the plethora of census data and other reliable secondary sources available in the United States, it is quite easy for a consumer goods company to identify the number of individuals or families in a given area that have the type of income, education, age, marital status, occupation, etc., that they see as potential customers for their products. If the potential market seems adequate in size, the company can then proceed to employ one or more market research techniques to allow them to discern the purchasing habits, needs, and wants of the particular group. If the company so desires, it can tailor its products to this target market's idiosyncratic consumer preferences. Again, the whole process of tightly defining the parameters of this target market is referred to as segmenting the market, while the act of tailoring the product to the buyers' desires is consistent with demand-side marketing and the widely held marketing concept that the consumer is king.

Segmentation for a Consumer Product

A national dairy may be contemplating the introduction of cottage cheese to its product line in the New England area. Either through experience with other dairy products or through the recognition that cottage cheese is not a basic food item, it may see its possible target market as being families in the middle to upper-income bracket. A review of census data indicates that a sufficient number of families in this income category are located in New England; and in addition, information on age, occupation, and education of that income group in New England is also provided. The national dairy can next conduct a survey in New England among families with these identified demographic characteristics and learn whether or not they eat cottage cheese and, if so, their present brand preferences, shopping behavior, and other pertinent information. An analysis of this data may then imply that a large market for cottage cheese exists and, in addition, that consumers' preference for cottage cheese with chives is not being met by existing dairies. Thus, the national dairy has not only identified a potential target market (i.e., customers tightly identified by a whole range of demographics and customer attributes), but it also has become aware of an unfilled market need.

Industrial Market Segmentation

Similar scenarios are repeated daily in the United States and in some developed overseas markets, as firms search for unfilled niches in existing markets or consider product or service modifications that will increase their market share. While the preceding illustration was a consumer product, a similar industrial (or business-to-business) example could have been used. Take the case of a company identifying its target market in the medical instrumentation industry for one of its products, a blood gas analyzer. Data on hospitals could be obtained from various American Hospital Association directories that would permit a firm to delimit its potential target market by region and bed size. A survey could then be conducted by the company (or a professional marketing research firm representing it) among key department heads in the selected hospitals to determine their needs for such equipment. The resultant information would provide a list of high potential customers for the firm's target market.

What Do the Dairy and Instrumentation Firm Have in Common?

First, the nature of the products in question are not really new; rather, in both instances we selected products whose need had already been recognized. Cottage cheese was well accepted in the New England market and was only new to the manufacturer. While the firm's blood gas analyzer could have many special quality attributes, the general concept was not a radical departure from the state-of-the-art. Second, the product usage and the general buyer qualifications or categories had already been determined. There was, in fact, precedence already established in terms of target market guidelines. Of course, the national dairy could sell its cottage cheese to restaurants or to the institutional market, and blood gas analyzers could be used in plants, military medical facilities, and so forth. There may be additional applications possible for both of these products, and the efforts in finding them might prove fruitful.

But, in reality, by following certain well-established segmentation guidelines, the two companies' target markets could be ascertained rather routinely. Assuredly, we knew the types of buyers and/or industry each represented, and in neither instance was it necessary to sell the concept before selling the product. As we see next, neither case offered us the challenge frequently provided by a high-technology product, service, or process.

Positioning: The Selection of a High-Technology Target Market

To make our point, it would be tempting to suggest that all high-technology products, services, or processes represent totally new concepts, are unheard-of breakthroughs, or at least through illustrations, we could treat them as such.

But, our own research and experience among high-tech firms suggest that many additions to their lines represent but modifications and improvements of existing products or processes, many of which are suggested by their customers or their distributors. Indeed, in a recent survey we found the following ranking of sources of new product ideas:

1. Customer feedback (requested)
2. R&D
3. Customer feedback (unsolicited)

While the remainder of the rankings are presented in chapter 5, it is interesting for our present purposes to note that R&D is not the high-tech firm's primary source of new product ideas.

Still, there are sufficient numbers of totally new products or processes to warrant our recognizing the need for a different approach to target marketing. What is desirable is a method that is peculiar to high-technology products. Further, we recommend following this different avenue to target marketing not only at the outset for the new products or processes, but also for all high-technology products until they gain routine acceptance in each of their priority areas of use or application. As their markets mature, of course, the traditional market segmentation activities become more appropriate.

What Are the Special Targeting Problems of the High-Technology Firm?

As we have described segmentation activities, it has most likely become apparent to the reader that we have been describing a market situation in which the producer can usually identify the appropriate industry or the broad customer group for its product, service, or process. Further, the uses for the product, service, or process are generally either inherent in its design and/or already have a degree of acceptance in the market. For example, if the Hoover Company were to develop a laser vacuum cleaner, the general purposes of a vacuum cleaner would already be manifest to its potential customers. There are exceptions, of course, such as the many uses of Arm and Hammer Baking Powder (i.e., everything from cleaning a white wall tire to baking), but these are rare.

New products, services, or processes do come on the scene. If it is a totally new consumer product, maybe the first home video recorder or hand-held calculator, efforts need to be made to stimulate a primary demand for the product, as well as selective demand for the particular brand or manufacturer. In these two illustrations, Sony and Texas Instruments played an important product information or education role, as well as a brand preference or sales role. However, the product uses were inherent in each of these products, and the costs alone designated what one of the parameters of the target market would be—middle- and upper-income consumers.

But what about the high-technology product or process whose applications are unknown or have not been fully identified? And what about those instances in which a product or process may have a market potential that crosses industry categories and thus makes it necessary to establish priorities for industries before attempting to classify the priorities into marketing subsets? These are not the only target marketing problems found in high technology.

Other problems relate to the speed with which the industry or product/process category may be developing and often to the concomitant reluctance of buyers to respond (buyers prefer to wait for later generations of the product/process). Also, some high-technology producers must choose the target market for a fraction of a new product development or process while they withhold the total breakthrough for competitive or profit-related reasons.

The list of targeting and other marketing problems unique to high-technology firms could go on. But a more relevant question concerns whether or not these problems are important enough to suggest the need for a different approach to segmenting a market, and thus, to identifying a desirable target market. We believe the answer is yes and believe that current battles being fought in the genetic engineering and robotics fields provide useful cases in point.

In the genetic engineering field, the initial skirmish has been conducted in the process patent arena. It has not really reached the applications phase yet. But the British scientists who did the basic "hybridoma work did not patent [it] because they were told it had no commercial potential."[3] Today the potential applications of genetic engineering are mind-boggling in the agribusiness industry alone. One nursery in California is reported to have the ability to clone a million plants in six months and all virtually disease-free—a revolutionizing of the nursery business.[4] Similarly, in the future, scientists see this industry producing everything from microorganisms that clean up oil spills to new forms of cancer therapy.[5]

The game will be won in the biotechnology industry just as much through the ability to recognize the most viable applications as through R&D itself. As the chairman of the board and president of Schering-Plough said in his annual message to shareholders on February 22, 1983, it is essential to create a technological advantage and "then convert it into a commercial advantage."[6] Traditional approaches to target marketing are not designed to accommodate this kind of marketing environment, where both rapidly changing competitor research successes and the potential applications seemingly are endless.

A somewhat different situation exists in the robotics field, although many of the same characteristics are present. This high-technology research field probably has been spurred on by public interest in "Star Wars," while simultaneously being viewed as the answer to U.S. productivity improvement and as a growing villain in the escalation of structural unemployment. Although most experts believe the robotics field is in its genesis, the number of competitors is staggering, with estimates of two hundred producers in Japan alone.[7] Undoubtedly, three important happenings will occur in the industry in the next decade: (1) consolidation (mergers and acquisitions) and a sorting out of the competitive milieu;[8] (2) striking new developments in manipulators which will rapidly increase potential applications; and (3) an increasing marketing sophistication. As an expert in the field suggested to one of us at the international robotics trade show held in Chicago in April 1983, "there are over one hundred robotics companies on display here today, while eight years ago we could have held the show in a closet." That demonstrates the increased level of competition in the field, especially when household names like Volvo, Siemens, Seiko, GE, and Westinghouse have joined in the battle with Ancumati, Milacron, DeVilbiss, and Hitachi. Based on predictions to date, the forecasted demand through the year 2000 will not support a fraction of this current competitive level. Therefore, the successful firms will be the ones that can expand applications, that is, identify and develop new target markets. Again, this is a common problem facing the typical high-technology industry.

What Do We Mean by Positioning?

In attempting to find the best approach to target marketing for the high-tech company, we have found many of its problems either to be singular or weighted much differently than those faced by more conventional marketers. We believe that market segmentation, as it has come to be used, does not adequately reflect the activities of the high-technology marketer toward the firm's target markets.

Stated simply, market segmentation is a too narrowly defined term to describe the target marketing activities that need to be employed by the high-tech company. Rather, positioning seems to best describe the steps that the high-tech marketer needs to follow if it is to identify correctly the firm's target markets and to place them in priorities.

Although market segmentation has come to be known as the managerial process of dividing a market or industry into manageable subsets, positioning connotes a degree of flexibility that allows the marketer to contemplate a broad range of users or applications for its product or process. The process of positioning allows the high-tech marketer to determine the various industries or buying groups that can use the products and array them in an ordered fashion. It is this emphasis on identifying applications that must be accounted for by any term used to describe target marketing by the high-technology company.

As a result of its positioning efforts, a high-technology company should be able to indicate which industries offer it greatest potential, which firms in those industries would benefit most from its products or processes, and which markets (in rank order) it will attempt to exploit. In our earlier discussion of the biotechnology field, we considered how Schering-Plough broadly described its target markets for the future. As we noted, there is a multitude of possible applications for the basic genetic technology the company could employ, but Schering-Plough has indicated it will pursue the pharmaceutical uses and build on its existing distribution network and experience. In another field—computer software—with an almost equally broad range of applications and industries to pursue, Cullinane first offered data retrieval systems running on IBM computers and then branched out to applications for banking.[9]

Both of these examples, however, are too broad to illustrate full significance of positioning. As the stages or steps in the following section will suggest, the identification of the target market for the high-tech firms should be just as specific as was the target market defined for the blood gas analyzer company referred to earlier in this chapter.

Positioning is a term that is consistent with the timing problems that are experienced by the high-tech marketer. The nature of many, if not most, high-technology products or processes suggest that the marketer has:

a limited product/process life cycle (i.e., the time required for the product or process to move from its introductory phase of acceptance until its demand ceases or production is halted)

a limited lead-time before competition responds with an equal or greater breakthrough or improvement

a desire to control the portions of its new technology that it introduces during any single time period or to any single target market

Most high-technology firms would delight in orchestrating the timing controls employed by Dr. Edwin Land as he patiently controlled the developments relating to the Polaroid camera. But such a successful positioning case is less likely to be repeated today. There are simply too many companies jockeying for the lead in most high-tech industries.

A recent description in *Business Marketing* of the rapid evolution that has occurred in the CAD/CAM field in the past five years is just one of the many illustrations of this timing phenomenon in high-tech industries. The crucial point is that "the technological edge in CAD/CAM keeps jumping back and forth between vendors," with no one firm in CAD/CAM having the luxury of a long lead-time at its disposal.[10] This rapid change puts enormous pressure on marketing, as well as R&D, and emphasizes that the correct selection and the establishment of priorities for target markets is essential, if any fractionalizing of new technology is to be achieved.

Stages in Positioning for the High-Technology Marketer

We have emphasized the need for management to identify all the alternative applications for its new products and processes. Undoubtedly, companies that have pioneered research in the various high-technology areas, rarely, if ever, actually produce and sell all the possible applications that may result from a new product or process. However, we are concerned that the firm have the go or no-go decision regarding the possible applications and that it identify these potential uses rather than missing fruitful opportunities by being unaware that they exist.

By formally or informally proceeding through the stages of positioning as shown in figure 3–1, the company should be able to identify most of the

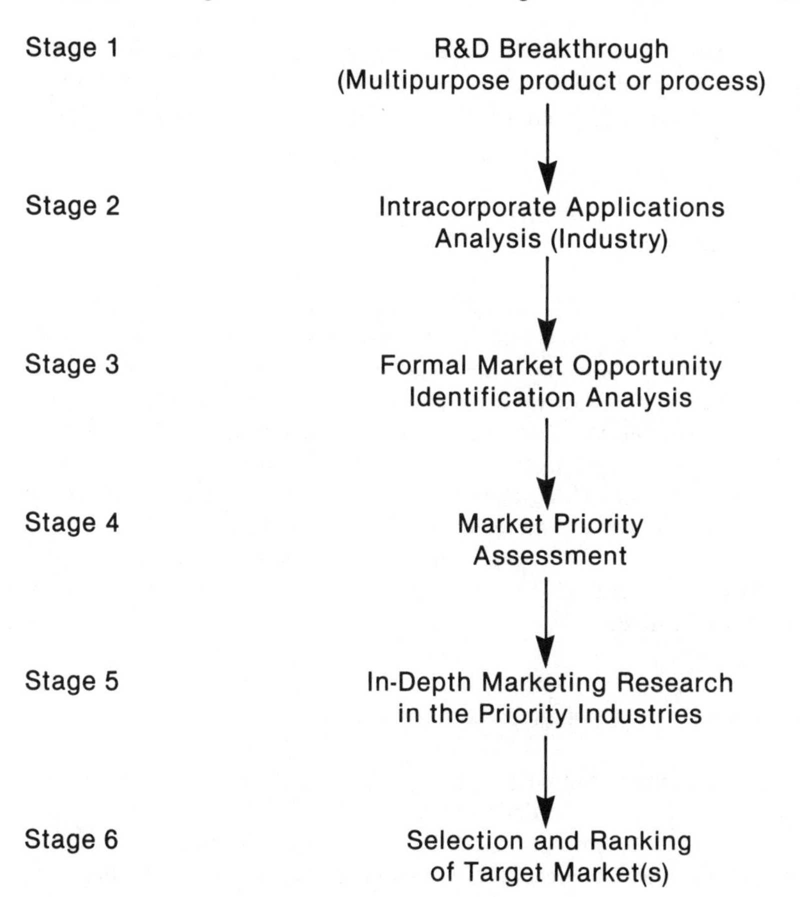

Figure 3–1. Stages in Positioning for Targeting High-Tech Markets

opportunities and potential target markets for its products or processes. In discussing high-technology marketing, of course, we constantly find ourselves in the old chicken-versus-the-egg dilemma. There are research breakthroughs made daily by universities, nonprofit and commercial research companies, in-house R&D staffs, and so on. Some are made with little or no concern for future industrial or consumer applications, but the bulk of breakthroughs is likely to be developed by companies or individuals seeking a solution to specific problems. This fact is very apparent in the medical-related field, where there have been recent efforts to find cures for diseases ranging from cancer to herpes.

However, we shall disregard this issue for the moment, since many products or processes may have even broader application than was originally intended. Instead, we will discuss the six stages in positioning that we believe a high-technology company needs to go through if it is to most effectively identify or target its market and to ensure that it does not miss a profitable application.

Stage 1: R&D Breakthrough

The first stage, which needs little amplification, merely suggests that a new product or process breakthrough has occurred. As is often true in areas like industrial robotics, computer chips, optical communications, genetic engineering, and computer software, the product or process probably has multiple uses. No matter whether the research represented a relatively pure breakthrough (the initial hybridoma or gene-splicing work) or whether it was an outgrowth of seeking answers to one or a few specific applications problems, it is essential for the firm to (1) recognize from the outset that the product has multiple uses and (2) identify what these uses are. Thus, the internal announcement of the new product or process should trigger the action described in stage 2.

Stage 2: Intracorporate Applications Analysis

This stage is designed to identify—based primarily on internal analysis—the potential applications. In order to ensure that this important stage occurs, we suggest that a formal committee be established, although our research with the robotics, electronics, and biomedical firms indicates that this is not now a pervasive practice. Involvement in this stage should include executive-level representatives from top management, R&D, and marketing.

In some instances, corporate policy may have given clear direction. For instance, top management at Schering-Plough has made a distinction between nearer-term, current-technology products and innovative longer-term projects that focus on applications relating to its pharmaceutical operations. Thus, Schering-Plough would focus on applications of hybridoma technology that relate to the pharmacists' shelves,[11] while, as noted earlier, the range of gene-splicing applications seems endless and cross many potential industries. To illustrate this even further, both Novo Industri of Denmark and Schering-Plough have a relationship with Biogen S.A. of France, an international biotechnology company. However, Novo Industri's research agreement with the Biogen relates to human insulin.[12]

The purpose of this activity is to take a first step in identifying possible applications of the new product or process. And equally important, the positioning team will decide whether there are possible applications in other, less-familiar industries than the one(s) on which the company has been focusing its attention. Brainstorming is just one technique that might be explored by the team. Another is to conduct focus groups among technical experts in the prospective industries to suggest possible applications that the internal review might overlook. (Chapter 4 offers a discussion of focus group analysis.)

Stage 3: Formal Market Opportunity Identification Analysis

The future opportunities in the identified industries should be explored more formally. This step involves extensive secondary research to determine the potential of each of the industries. Much of the data analyzed is quite similar to the financial and market information that would be reviewed in considering an acquisition or merger with a firm in that industry. Included would be a review of market forecasts, demographic patterns, average ROI figures for the industry, and the like. For example, assume a biotechnology breakthrough in the genetic engineering field has potential applications for the agribusiness industry. This possibility would necessitate an assessment of the long-term potential in that industry, including the forecasted future demand for select agribusiness products, the ROI and/or return on shareholders' equity picture of agribusiness firms, the level of competition in that industry, etc.

After completing this stage, the positioning team will have identified several possible applications for the new product or process that will cut across industries and, in addition, the team will have information on the future prospects for these industries. Some applications or industries can probably be eliminated from further consideration. These eliminations may be due to the long-term prospects of the industry, to the level of competition in the industry, to the difficulty associated with the development of applications, or for other reasons.

Stage 4: Market Priority Assessment

The industries and related applications that remain after the review in stage 3 would all have some potential. Realistically, however, the company may not be able to exploit fully all the potential applications that have now been identified, even if all of the industries prove to have exceptional prospects. In fact, such an action may not be wise, since the timing for a move into some industries might not be right.

However, before doing additional marketing research to target opportunities better, it is important to initially do some ordering of or establishing priorities for the industries. The cost alone of conducting extensive marketing research necessitates first identifying the industry or industries that appear to offer the most profitable opportunities.

Stage 5: In-Depth Marketing Research

Among the remaining highest- potential industries, the company can then conduct research to detail more thoroughly the nature of the customers in each. In business-to-business marketing, the members of any industry can undoubtedly be classified by size, profitability, and other objective characteristics, and may be grouped subjectively by considerations such as innovativeness or leaders and followers.

Primary (original) research can be carried out to determine which of the firms are interested in the application of the new product or process to their industry. More specific demand information can be obtained and a better forecast of the opportunities prepared. Moreover, the special needs of the various members of the industry can be assessed. This evaluation may require modification of the firm's product or process, if particular potential customers are to be satisfied. A company may uncover additional uses for its product or process that would allow it to tap niches in the industry that are not currently being filled by the competition (substitute products or processes). In endeavoring to gain additional information or uses, some companies have even found it useful in this stage to allow key innovative firms in the industry to try prototypes of the product or process. Also, trade shows are used widely to obtain feedback regarding the likely success of new products or processes in the appropriate industry.

It is in this stage that the more traditional segmenting of the market (i.e., identification of subgroups) occurs.

Stage 6: Selection and Ranking of Target Markets

The capstone stage of the positioning process involves the selection of the specific target market(s) for the product or process within the priority industry (or industries). Many factors come into play as the positioning group sifts through the inputs obtained in stage 5. Some of the factors are:

the potential demand provided by the various segments of the industry

the accessibility of the various segments in terms of location the state-of-the-art level in the various segments (a technological breakthrough may be too extreme for some)

the vulnerability of various segments (some may have greater need of the product or process than others)

As noted earlier, some immediate niches may have been identified where some modification in the product or process could enable one to have virtually no competition.

The final decision of the positioning group does involve other considerations. For example, does the company wish to play a leader or follower role, that is, be the innovator in some segment or await its more full development? Or do some applications represent diversions from the company's main strengths? Next, could the firm license its product or process to allow someone else to tap a coveted segment and still reach other segments through its own production and marketing efforts? All such considerations are relevant as the team arrives at the final identification of its immediate target market(s). For a particular software producer, as an example, this could mean that it has chosen to focus on the banking industry for its new application, and more specifically, sees its primary target market within the industry to be smaller banks (for example, those with assets of x number of dollars or less, located in the northeastern region of the United States).

At the end of this stage, the positioning team or group of decision-makers' efforts for this product or process innovation are complete, as the target market or markets have been identified. The new product development techniques, described in chapter 5, now come into play.

Market Segmentation and Target Market Refinement

Are the more traditional market segmentation activities appropriate as the new process or product applications become accepted and the markets become more fully developed? The answer is yes. Our focus until this point has been on the breakthrough product or process, and we have deliberately stressed the need for the firm to try to ensure that no possible application is overlooked. This perspective is very much a supply-side marketing orientation and emphasis. Our examples have been taken from some of the fastest-paced supply-side fields; one need only attempt to follow developments or the patent race in genetic engineering to see the importance for positioning efforts in arriving at target markets.

This level of activity characterizes many of the glamour fields of high technology—robotics, lasers, fiber optics, and biotechnology. However, as the definitions in chapter 1 denote (pp. 18–19), there are many other fields and industries that also fit the high-tech nomenclature.

In discussing the marketing of high-tech products and processes, we must accommodate a broad range of company needs and problems. The types of firms differ widely. As pointed out in chapter 2, they range from large firms totally in high technology, like Texas Instruments, Perkins Elmer, and Atari, that have some products which have reached maturity or have been pruned, to very small high-tech companies similar to Insystems, Inc., and Spire, Inc., which are caught up in highly competitive fields. For example, *Inc.* suggests that computers and related products were the second fastest growth area for its *Inc.* 100 between 1978 and 1982.[13] All of these firms must be concerned with market segmentation and the resultant target market(s), in addition to the positioning process.

Market Segmentation

We now take a slightly different look at the process of market segmentation. Simon Majaro, an English author, offers a very useful definition of market segmentation. He says it "is a strategy that enables the firm to maximize the results of a given marketing effort by exploiting clearly identified strengths in relation to a submarket which is either inadequately satisfied by other manufacturers or where the firm is particularly well placed to do an effective job.[14] What is important in this definition is that it reflects a down-to-earth perspective; it does not imply that the product or process is a breakthrough, but instead recognizes that at some point it must begin to battle for its share of the market along with other products from firms having similar objectives.

This view is most realistic, as it signals to us that the edge gained by the new product or process often starts to wane after introduction. At the very time that revenues begin to grow as a result of its carefully positioned entry, the company finds itself moving from being a supply-side marketer to a demand-side marketer. The marketing vice-president or manager now has full responsibility for the new product, and the concerns of the marketplace become preeminent.

In our previous section, one dynamic of the high-technology field mentioned was the difference in size of firm. The large multiproduct high-tech producer, a Texas Instruments or Schering-Plough, will have products or processes in various stages in the life cycle of a product. (The product life cycle concept is discussed more fully in chapter 5.) For a new therapeutic entity, possibly recombinant DNA-related, a major pharmaceutical firm may act as a supply-sider, yet it has to be flexible enough to be marketing a host of other pharmaceutical products that may have long since lost this degree of glamour and are consequently in mature or even declining positions in terms of the product life cycle. Mature or maturing products or processes require marketers to have an increasing understanding of the marketplace or, in short, to become more adept at market segmentation. Smaller firms that specialize in high-technology products or processes face a similar or greater challenge as they have products or processes in the different life-cycle stages. They have fewer products to carry the profit-producing load as new innovations or breakthroughs are developed. They, too, must become highly adept at market segmentation. Their task is to find out more about the needs of existing customers, in endeavoring to retain them, while simultaneously seeking new target markets.

What Are the Ways to Segment a Market?

The bases or ways for segmentating markets are virtually limitless and are confined only by the imagingation of the marketing group. The selection, of course, is restricted by factors such as the amount of competition, the nature of the product or process, the life-cycle stage, the extent of differentiation from competition, and smaller concerns.

Most striking is the degree of similarity between the bases of segmentation that can be employed for consumer goods and industrial products or processes. For example, an industrial market can be segmented into groups according to the companies' annual revenues, while consumer markets can be classified according to family or personal income. Likewise, markets for consumer goods and industrial products can both be classified by age (personal versus corporate); size (family versus number of employees); price sensitivity (generic product consumer versus low-bid buyer); method of purchase (credit, lease, or cash); region of the country (same for both); and innovativeness (innovator or follower for both). Still, there are essential differences between segmenting consumer goods technical product/process markets and their industrial and government technical market counterparts.

Consumer Goods Markets. No phase of marketing has been more carefully studied than consumer behavior. While we are concerned with generalities here, publications such as the *Journal of Consumer Research,* the *Journal of Marketing Research,* and the *Journal of Marketing* are filled with articles describing more sophisticated consumer research for those seeking additional information. However, the studies these kinds of journal report rarely involve high-technology products, and thus, their importance relates more to methodology than to industry- or product-specific findings.

The consumer goods marketer normally begins his or her market segmentation efforts by using readily available demographic data from secondary sources (U.S. Bureau of the Census, etc.). And with a few assumptions regarding purchasing behavior differences based on age, education, income, location, and so forth, this may prove sufficient for making a "first cut" at segmentation. In the introductory stages of revolutionary high-tech products, say color televisions or microwave ovens, a very useful first step would have been to divide the market by age, education, and income.

However, in later stages of the markets, demographic factors are not likely to "be precise enough measures of buying preferences."[15] Therefore, companies turn to psychographic (or life-style) segmentation, benefit segmentation, a combination of the two, or some other type of psychological or sociological research. Psychographic segmentation explores behaviorally oriented facets of buyer behavior (for instance, personal interests or hobbies, political and social orientation) and permits the development of consumer life-style profiles. These market segmenting efforts require the company, usually through an outside marketing research firm, to conduct the needed research, which is expensive. However, expensive or not, it can be quite useful in targeting specific groups and directing strategy toward them. Benefit segmentation requires the firm to commission or conduct marketing research to identify specific markets based on the benefits different groups of customers receive or hope to obtain from this type of product. Benefit segmentation involves surveying potential buyer segments and/or conducting focus groups to identify very specific product needs or desires.

By combining demographic segmentation work with benefit segmentation or psychographic data, the company can have a pretty good idea of its prospective customers. This market definition allows the company to modify its product more effectively, select its advertising message and media, determine its distribution approach, and price its product.

Industrial or Business-to-Business Customers. Paralleling the demographic information available on the consumer market is the extensive secondary data readily obtainable on industrial or business-to-business markets. Besides the U.S. Census of Business data, there is a host of information that is available from trade associations and private organizations, notably *Moody's, Standard and Poor's,* and *Dun and Bradstreet.* It is possible to obtain everything from sales and profit data to the names of the individual firm's officers. This information offers a variety of ways to classify prospective buyers for market segmentation purposes.

At the same time, many of the buyer behavior approaches that have been employed in consumer marketing have begun to be employed effectively in industrial markets. For example, a company can conduct focus groups or surveys among its prospective customers or customers' purchasing managers or engineers to learn more about what a type of product is expected to do (i.e., a benefits analysis). Or a company may use the same methodologies to explore the interests and reading, listening, and/or viewing habits of a select level of company officers in an effort to determine where to place advertising and what theme to employ. In other words, they should conduct a form of life-style research and determine a profile of the chief decision-maker or influence in their target market. One thing is clear about industrial market segmentation—it is now recognized that it is helpful to go far beyond demographic segmentation in effectively targeting markets. We also believe that the characteristics of many high-technology business-to-business customers, often better educated and more entrepreneurial, make more sophisticated marketing research necessary.

Company Image. Finally, larger high-technology companies frequently carry out image research to learn how they are perceived by their customers, prospective customers, community, and for national leaders. They are usually concerned with how the company is viewed and what its awareness level is relative to the competition. To counteract low awareness or poorer image totals, firms often subsequently employ corporate advertising campaigns. (For more details on advertising, see chapter 6.)

In our judgment, strong image or awareness scores can be used as a basis of market segmentation, when included with the other bases for segmentation. And this is true for either consumer or industrial/business-to-business market segmentation. Assume that a high-technology firm targeting on consumer markets has conducted both demographic and psychographic segmentation and, employing these inputs, has carefully identified its market segment. By using its image research inputs, it finds that its strongest image and awareness scores are found in the Midwest. It can then focus most heavily on that fraction of its total target market area, while using corporate advertising in the other markets.

Global Realities

High-technology marketing is a global activity, and positioning and market segmentation should likewise be viewed as global activities. If one doubts this, he or she need only visit a robotics trade show where producers from Japan, Sweden, West Germany, France, the United States, and other countries come to exhibit and sell.

Consequently, in positioning its products or processes through segmenting and targeting its markets, the company needs to consider both domestic and nondomestic opportunities. It should recognize and exploit those world areas where state-of-the-art differences may make a product or process that is maturing in the United States or France remain in the growth stage there. Yet, for many products or processes, the gap between markets is minimal today. This reality means that a company needs to exploit its application to a worldwide industry.

Likewise, a gap or niche that has been found in the market for a particular product in a given industry often is a global gap. This opportunity permits the firm to reach scale economies—a level it could not attain if it were marketing it in a single country.

We raise these points because U.S. industry in general has often been correctly charged with being too domestic-oriented. Of course, a few high-technology fields have been prevented by the U.S. government from marketing in politically sensitive markets for national security reasons. This reason, however, cannot be used to explain plausibly why many other high-technology companies have taken a domestic-only view of their potential opportunities.

Notes

1. Philip Kotler, *Marketing Management,* 4th ed. (Englewood Cliffs, N.J.: Prentice-Hall, Inc., 1980), p. 82.
2. Levitt, p. 94.
3. Tamar Lewin, "The Patent Race in Gene-Splicing," *New York Times,* August 29, 1982, p. F-4.
4. Richard C. Davids, "Here Comes the Next Green Revolution," *World,* Number 4, 1982, p. 6.
5. Lewin, p. P-4.
6. Richard J. Bennett and Robert P. Luciano, "Letter to Shareholders," *Schering-Plough Annual Report 1982,* p. 3.
7. "Robots Bump into a Glutted Market," *Business Week,* April 4, 1983, p. 45.
8. Richard A. Shaffer, "Markets for Robots Turns Sour, May Speed Industry Shakeout," *Wall Street Journal,* April 22, 1983, p. 21.
9. Andrew Pollcak, "Computers: The Actions in Software," *New York Times,* November 8, 1981, p. F-1.
10. Philip Maher, "CAD/CAM Vendors Plot a New Course," *Business Marketing,* April 1983, p. 67.
11. "Schering-Plough and Biotechnology," *Schering-Plough Annual Report 1982,* p. 8.
12. *Novo Annual Report 1981,* p. 8.
13. Donna Sammons, "The Inc. 100," *Inc.,* May 1983, p. 51.
14. Simon Majoro, *International Marketing: A Strategic Approach to World Markets,* rev. ed. (London: George Allen and Unwin, 1982), p. 42.
15. Richard T. Hise, Peter L. Gillette, and John K. Ryans, Jr., *Basic Marketing: Concepts and Decisions* (Cambridge, Mass.: Winthrop Publishers, Inc., 1979), p. 46.

4 Searching for Market Opportunities and Insight

No product idea has a manifest destiny that will propel it to market success.

The world of high technology is curiously dichotomous, and this dichotomy makes the marketing task extraordinarily challenging. From the producer's or supply side of the market, there is emphasis on and keen understanding of the latest in technology. Yet, on the other side of the market—the buyer's or demand side—there often is a widespread lack of knowledge and understanding of science, mathematics, and technology. To be sure, there are exceptions, notably in certain business-to-business marketing situations in which both the seller and buyer are technologically adept. But generally, the technological interest and acumen of the seller far exceed that of the buyer. This imbalance can be seen even when the buyer has technological education and training.

Is it the buyer's place to improve his or her knowledge of the seller's technology? Hardly. In this imbalance between what the would-be seller of technology knows and what the potential buyer of technology does not, it is the latter that must prevail and control. Because of the buyer's prerogative to say no or to look to alternative ways or suppliers to satisfy a need or want, it is the task of the technologically oriented seller to explain and persuade of the benefits of purchasing or using the product. The seller cannot count on the prospective customer to seek out how a technology, especially one that the customer does not understand, might be of benefit to him or her.

No product idea has a manifest destiny leading to market success. The adage "If you build a better mousetrap, the world will beat a path to your door" is wrong. To see why, take the adage quite literally. An elaborate electronic mousetrap, with sensors to identify the prey and spring the trap on it and with chimes that gently alert the trap's owner that a kill has been achieved, is not assured of even a modicum of market acceptance. By any technological measure, this electronic mousetrap is superior to the old cheese-wire-wood mousetrap. But a better mousetrap in the eyes of the buyer goes beyond what is better in a purely technological sense. The buyer is more familiar and comfortable with the existing technology—he or she understands it. And, besides, from the buyer's comparison of costs to benefits, the older-type mousetrap is a better perceived value. If the electronic mousetrap is to garner market success, it will be because its advocates convincingly explain the benefits and superiority over the old mousetrap to the largely nontechnologically oriented buyer. It would not be because the world rationally recognizes the obvious technological advantages of the new electronic mousetrap and beats a path to the door of the seller.

Like all organizations that intend to survive and prosper, high-tech firms must have their strategic compasses pointing outward to the market. They must be market-driven. But on this point there is a distinctive difference between high-tech firms and most other kinds of companies. The discernment of market opportunities in the form of buyer needs and wants is more difficult and problematic in existing or potential high-tech industries because buyers are often not aware of the technological capabilities that exist (or could exist) to solve their problems and fulfill their current or latent desires. For instance, were inexpensive, in-home, monoclonal, antibody-based tests for pregnancy developed and marketed because there was a ground swell among women of childbearing age asking for them? No, their market success was due to astute need discernment and technological application by a few biotechnology firms.

High-tech marketers deal with possible future demand that may not be readily discernible or predictable by surveying or studying present markets or market opportunities. In this sense, high-tech companies sometimes create new markets via the innovative adaptation and application of technology to problems or, occasionally, through the discovery of radically new technological innovations. Now one can argue that high-tech firms really do not create the demand, but rather, that the demand is latent and has been there all the time—the argument that there is nothing new under the sun. These debates are over semantics and, for practical purposes, are sterile. Even though the need for in-home pregnancy tests has obviously existed ever since human consciousness, there was not a clamor asking for the need to be fulfilled because women were not aware of the biotechnological possibilities that eventually made the tests feasible.

Potential buyers can normally provide useful product development and strategy direction to demand-side marketers. It is not difficult conceptually for consumers to evaluate automobile styles (even extreme automobile styles) for General Motors or for engineers to discuss the pros and cons of steel products. But many high-tech, supply-side firms exist in a world not all that conversant about technology. A potential buyer can be asked "what if you could . . . ?", but if this is beyond the realm of credibility, his or her answer has little validity. This perceptual and knowledge gap between potential buyer and seller makes the identification of market opportunities more complex and more tentative in high-tech industries. This, in turn, accounts for the relatively greater risk in high-tech companies. This high-risk kind of milieu goes hand in hand with innovative technology and is thus unavoidable. The fact is, those that pioneer and lead in most any venture are at greater risk than those that follow. Nevertheless, at least the chances of failure—of nonmarket acceptance—can be mitigated by the creative and selective application and tailoring of marketing research to the special problems of high-tech marketing.

Marketing Research: The Make or Buy Decision

How much should the high-tech firm's operating managers and technical staff involve themselves in marketing research? Is it best for them to defer predominantly to marketing research specialists inside and outside the firm?

Marketing research is far too essential to marketplace success to leave exclusively to experts in research methodology and techniques, regardless of whether they are outside suppliers or internal company consultants. Masahiko Kajitani, the general manager of video planning at Matsushita Electric Industrial Company, agrees with this assessment. He believes that a critical problem with U.S companies is that they too frequently delegate marketing and consumer studies to individuals not directly involved with developing and marketing products. As a consequence, invaluable personal dialogue between key company employees and customers fails to take place. Kajitani's own personnel *regularly* converse with store owners and make in-home visits to customers to ascertain what they like and do not like about Matsushita video offerings. Perceptive U.S. companies have now done an about-face or otherwise modified their thinking and are following the same tack. Design engineers at General Electric are being sent out to obtain customer opinions on GE's electronic wares. The president of the high-tech Intercolor Corporation, David Deans, has remarked, "Today we spend much more time talking to our marketing people and customers to find out what they—not the engineers—want."[1]

This direct-customer-contact approach for key company employees is needed to avoid the perils of their becoming isolated from customers. Employees not familiar with nor empathetic to customer problems and needs are prone to Rube Goldberg solutions and to overlooking opportunities. For technology and present or future market demand consistently to mesh in a timely fashion, the technical side of the business must be attuned to customers. First-hand experience with customer wants and difficulties is the best teacher. That is why marketing research is too important for the high-tech firm to totally farm out to anyone.

Does the importance of direct contact between a high-tech company's key people and its customers mean that there is little or no need for utilization of professional marketing research, especially from outside suppliers?

No, it means only that the marketing research task should not be left mainly or exclusively to specialists. There are, in fact, many kinds of marketing research that are best left to individuals with the requisite expertise, experience, and facilities. And, we might add, there are a number of start-up marketing research firms specializing in serving high-technology companies. These firms normally focus on particular industries—electronics is popular—and are liberally staffed with personnel technologically conversant with the relevant area.

Although marketing research should not be farmed out entirely, neither should the high-tech company undertake studies on its own that it does not have the internal expertise to conduct competently. A little knowledge is indeed dangerous. Countless times we have heard companies say they only intend to conduct a little survey or send out a short questionnaire or hold a group interview with customers. Many regret it later. In attempting to save dollars, they often end up costing themselves dearly when the results of their poorly designed and/or executed marketing research studies lead to marketing mistakes that usually could have been foreseen and averted.

Then, too, there is the issue of objectivity. Outside suppliers are more likely than the high-tech firm's own employees to carry out and report objective research. Being in the employ of a company builds in a bias—an unscientific predilection toward the company's ideas and products. Such favoritism is even more tempting for employees who are integrally part of the research, development, and formulation of marketing strategy for a product that is being evaluated for commercial feasibility.

This is illustrated by a case described here. A division of a *Fortune* 500 company was holding a freewheeling, in-depth group discussion with about thirteen of its customers. Over the objections of the marketing research staff, the divisional general manager had insisted on sitting in on the discussion in the guise of a customer. Part way through the session, criticism of the division's products became severe from a few of the customers participating in the discussion. The general manager lost his temper, identified himself to the group, argued with the critics, and thwarted valuable feedback. Even though this is an extreme and probably rare case, it does nonetheless point up the lack of objectivity that is usually part and parcel of high involvement with a company, a product, or anything else.

Our counsel is this: Heavily engage key high-tech company employees in customer research, especially the often-isolated R&D people. Getting them regularly and systematically involved in personal discussions with customers about the latters' needs and problems is vital. If this interaction is not provided for systematically and regularly, it will not achieve the desired result. In the press of everyday business, such personal interaction will become less frequent. We also advise that outside suppliers be used, or internal marketing research capabilities, whenever more sophisticated data

Marketing Research Manager for High-Technology Company

HT Systems, Inc. is a leader in a number of dynamic and exciting growth markets involving laser technology. In only ten years in existence, we have been fortunate enough to have attained sales that make us one of the largest 1,000 industrial corporations in the United States. Presently, we are also actively seeking market expansion abroad.

We are looking to employ a professional with 5 to 10 years marketing research experience in a high-tech environment to manage our corporate marketing research function. The position title is Corporate Manager of Marketing Research. HT Systems, Inc., is a decentralized, multidivisional company. The corporate marketing research office provides internal consulting to both corporate top management and to the various HT Systems divisions.

The primary requirements for this challenging position are the abilities to communicate effectively with both management and technical personnel and to guide R&D toward profitable market opportunities. The manager of marketing research is the main liaison between management and research and development. Generally, we are seeking a person with a proven track record in marketing research within high technology, who holds a relevant doctorate or an MBA. This individual presently might be the number two marketing research person in a high-tech firm or he or she may be a consultant with the proper credentials and experience.

If this opportunity interests you and you think you have the position qualifications, please write us in confidence. Include a current vita which includes salary history.

Mr. J. W. Porter
Corporate Manager of Human Resources
HT Systems, Inc.
1001 Old Castle Road
Austin, Texas 78745

The demand for marketing research within high-tech industries is growing by leaps and bounds. The HT Systems, Inc. ad, which is a composite, typifies the position listings that are appearing in major business publications. Via the *Wall Street Journal,* General Electric has looked for market research professionals to work at the very highest corporate levels on high-tech marketing issues and products. Candidate requirements included a doctorate in marketing/business and significant experience in industrial marketing. A multidivisional San Francisco high-tech company has sought a senior corporate level marketing research executive whose duties would entail strategic business planning involvement, managing information sources, and presentations to upper-echelon management. Applied DNA Systems of Pittsburgh, a technology transfer organization developing new biologically derived products with university scientists and contracting with major industrial firms for commercialization, was seeking an MBA with an appropriate technical undergraduate degree for market research and development. A defense systems company, specializing in computer services, has advertised for a graduate-level-trained and innovative marketing person to link systems technology to the strategic marketing process.

The growth of high-technology industries has also spawned the formation of marketing research firms focusing on the needs of specific industries. Some of these firms have achieved astronomical growth in terms of sales revenues and numbers of employees.

Market research in high-technology industries is increasing because of the now widely recognized need to direct R&D to existing or anticipated market opportunities. More and more companies are becoming market-driven.

---

collection and analysis is required. Jointly, this combination approach builds a healthy, open environment for decision making that leads to better results in the marketplace. More importantly, it gives the technical personnel a hands-on appreciation for market opportunities and requirements.

Diffusion of Technological Innovation: Basis of Insight into Marketing High Technology

The purpose of marketing research is to guide the company that undertakes it toward success in the marketplace. Marketing research is a way to provide the firm's executives with keen market insight; it offers a special understanding of buyer behavior that leads to marketing strategies that prove successful.

The accomplished marketing researcher's repertoire of methods and techniques for evaluating buyer behavior is extensive. His or her conceptualization, interpretation, and understanding of the problem (or issue) at hand determines what data are sought, and by which techniques, to gain insight about how the company needs to proceed. Thus, if the market researcher has a strong conceptual basis for thinking about problems to begin with, his or her research approaches have a better chance to be appropriate. For this reason, we know that the quality and appropriateness of marketing research conducted by and for high-tech companies depend largely on the researcher's understanding of just how high-tech market behavior works. We are of the opinion that, for marketing purposes, a good theoretical understanding of high-tech market behavior is more important than technical expertise in a particular technology, product, or process. Underlying technologies differ greatly across high-tech industries, but the fundamental principles of high-tech market behavior do not.

The code for understanding high-tech markets and marketing strategy is the concept of diffusion of technological innovation. It is the basis for developing useful high-tech market insights, for effectively researching high-tech markets, and for formulating consequent marketing strategies that prove capable of achieving the company's goals. Marketing research and strategy development for high-tech markets without a thorough familiarity and appreciation for diffusion of innovation is analogous to a psychiatrist diagnosing and prescribing without adequate grounding in the concepts and theories of human behavior.

The body of seminal and corroborating empirical evidence concerning diffusion of technological innovation is truly massive and, fortunately with few exceptions, has been found to be applicable to virtually any kind of genuine technological breakthrough. Numerous researchers from a variety of disciplines have contributed to the plethora of findings pertaining to diffusion of technological innovation. However, Everett M. Rogers is universally recognized as the pioneering and most influential individual involved. He is credited with development of the theory and model of adoption of innovation and with the location of diffusion research findings at the Diffusion Documents Center at Michigan State University. This evidence applies precisely to the types of significantly new products dealt with by high-tech marketers, rather than to the many so-called new products in mature industries that are often only minor variations of existing products. "New-improved" soaps or annual style changes in automobiles or appliances are not accommodated by diffusion theory. Microwave ovens, new generations of computers, unique biotechnic applications, and the like, are the proper genre of products and processes.

Figure 4–1 depicts how an innovation diffuses or spreads out over time. The vertical axis shows the innovation gradually achieving additional market acceptance until it plateaus at a certain percent of the total market potential. For some products, that percentage might approach one hundred (color television sets in U.S. households), whereas for other innovations (home movie projectors), the percentage would be much lower. The horizontal axis depicts time. Some innovations might take one hundred years or more to attain their market saturation point (maybe nuclear power plants) and other products could do so in a relatively short period of time (perhaps a generation of personal in-home computers).

A more illuminating behavioral look at diffusion of technological innovation is illustrated in figure 4–2. Countless empirical studies support the notion that most technological innovations diffuse through the market in the manner of a normal distribution, as traced by a bell-shaped curve. Not all buyers who eventually adopt an innovation do so at the same time. (Adoption means the firm decision by a buyer to use an innovation regularly.) Some people or businesses adopt a product more quickly than others,

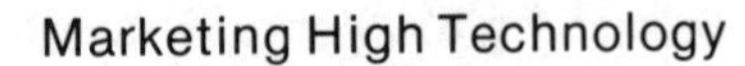

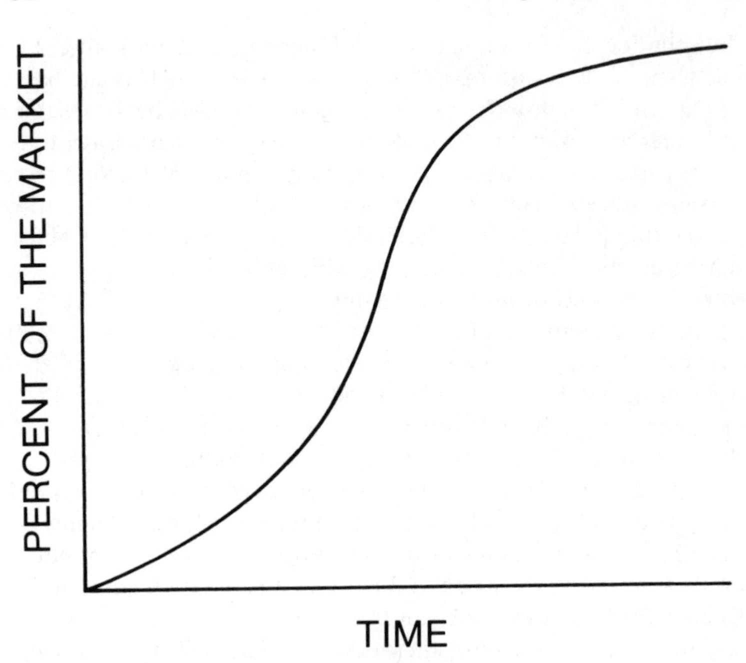

Figure 4–1. The Diffusion Process

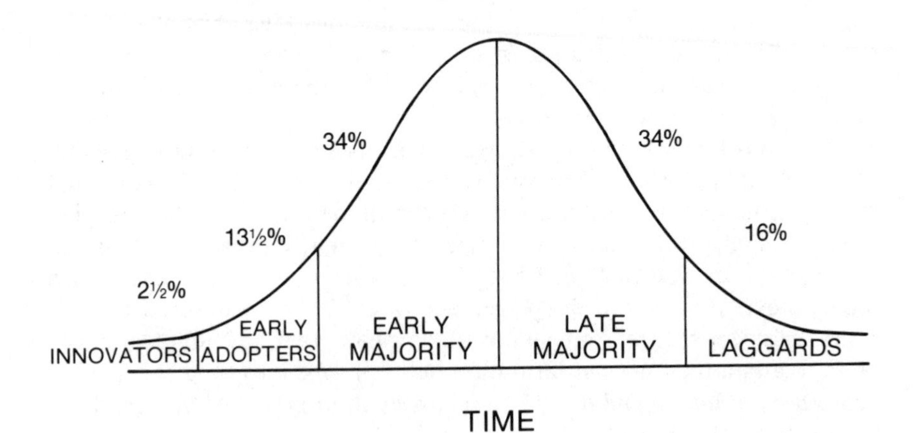

Figure 4–2. A More Behavioral Look at the Diffusion Process

and some never adopt it. On the basis of when they adopt, buyers can be classified into categories as shown in figure 4–2. For purposes of analysis, the initial 2½ percent of the buyers to adopt an innovation are aptly named innovators, the next 13½ percent are labeled early adopters, and so on through the final 16½ percent of the buyers to embrace an innovation, who are understandably termed laggards. Again, the diffusion time that lapses from innovators through laggards can be brief or long. The essential point is that there are significant differences among buyers in the various categories that can be meaningful for high-tech marketing strategy. These variations can be used to increase the absolute number of buyers who ultimately adopt an innovation as their own and to speed the time it takes for the innovation to run its full bell-shaped course.

For conciseness, the term *innovator* is used here to designate buyers who tend to purchase products early in the product life cycle. Thus, this definition encompasses what are referred to in figure 4–2 as innovators and early adopters. While there are differences between innovators and early adopters, the combining of the two categories does not cause significant analytical problems.

When compared to the other categories of buyers—the early and late majorities and the laggards—innovators have been found to have different social, economic, and psychological characteristics. In general, innovative buyers are more:

respected by others and, as a result, are opinion leaders (not in all areas: only in matters in which the innovator/opinion leader is seen as having competence)

favorably disposed toward change

willing to assume risk and venturesome

socially aggressive and extroverted

communicative

productive, as indicated by their higher incomes

cosmopolitan

knowledgeable about what is going on in the world

exposed to mass media

in touch with change agents, such as sales representatives

Later in this chapter, we look at how a high-tech company might go about identifying innovators for use in product development and in actual product commercialization. For now, the aim is to depict the vital role that innovators can and do play in determining whether or not a high-tech product or process becomes a market success and, if successful, how quickly success is achieved.

Innovators are instrumental because they are respected by others as being knowledgeable and truthful about a class of products or in a special problem area. Consequently, an opinion leader is one whom others listen to or come to for advice whenever a problem or decision falls within the opinion leader's sphere of knowledge. The opinion leader can have an enormous influence on risk-averse buyers. And, importantly, innovators are not beyond the reach or influence of well-designed marketing strategy. By this we mean that innovators can become central elements in a company's marketing strategy. Properly accounted for, innovators are the best risk-reducers available to the high-tech company as it goes about trying to develop and market products that gain widespread and rapid acceptance.

Besides using innovators, what else has diffusion research indicated that high-tech companies need to do to improve their odds of success in the marketplace? Studies in diffusion of technological innovation point to a number of helpful marketing-related generalizations that pertain to the way an innovation is presented and communicated to potential buyers, and hence perceived by them. These generalizations are embodied in several questions that the high-tech marketer needs to ask and answer satisfactorily before bringing a product to the market for the first time or giving up on a product that is already on the market but has not caught on.

What are the advantages of our innovation vis-à-vis alternative means of fulfilling the targeted buyers' needs and wants? Stated differently, what are the immediate benefits of our innovation relative to the competition? From a cost-benefit viewpoint, how will the prospective buyer contrast our innovation (the electronic mousetrap) to competitive products (the wood-wire-cheese trap)?

How can we persuade intended buyers that our innovation will not be disruptive—that it is in fact compatible with existing lifestyles (microwave ovens for preparing meals) or work styles (word processing equipment for preparing correspondence)?

How can we reduce the complexity of the innovation as perceived not by us, but rather, by the potential buyer? We are thoroughly familiar with our innovation and the intended buyer is not. Can we somehow reduce the buyer's perceived risk, skepticism, reluctance to change, and possibly fear of the innovation, by making it divisible so that the potential customer can try it, or lease it, or return it before making an irrevocable commitment to adopt it?

Can we clearly and convincingly communicate our case to our target audience of prospective buyers? Have we taken steps to do so?

The ultimate degree of market success enjoyed by a high-tech product and the speed with which success is achieved hinge largely on management's ability to come up with the right answers to this set of questions.

From the massive body of data and information derived from studies of technological diffusion of innovation, a few imperatives for high-tech marketers are encountered time and again. They are:

When researching and developing the innovation, deeply involve potential customers, especially a disproportionate share of innovators/opinion leaders.

Even though the technology underlying the innovation may be complex, keep the innovation as simple as possible for the intended buyer to use.

Whenever possible, focus the initial marketing efforts on innovators/opinion leaders.

Conduct face-to-face demonstrations of the innovation for prospective buyers if feasible. If not practical, demonstrate vicariously through advertisements, particularly in the electronic media where visual effects are best. Better still, let the prospect conduct his or her own demonstration and trial.

Take care to educate potential buyers in other ways also, so that they can evaluate the innovation better and see how it immediately benefits their life styles or work styles, and so that their fears are overcome.

Creating and Considering New Product Ideas

Rapidly evolving market conditions directly attributable to technological progress mean that the lifeblood of high-technology companies is new products. A high-technology firm either goes along with the flow of new products by innovating or quickly following with its own, or it perishes in obsolescence. These are the only choices. And, because no company is given to intentionally committing corporate suicide, the creative and analytical process of generating and weighing new product ideas is one that needs to assume a great deal of management's time and effort.

Revolutionary ways of doing things, the bona fide technological breakthroughs, do not come along often in the course of history. More typical are incremental improvements in existing technology and the creative application of existing technology to buyer needs and wants. The ways that existent technology can be applied to solving problems are usually plentiful. Ponder, for instance, the number of possibilities yet untapped for bringing robotics and biotechnology to bear on human needs. From the almost endless assortment of market possibilities before it, the high-tech company must select those with the most potential for achieving corporate objectives. It is in this priority-selection task that marketing research is so fundamental and necessary; it's the very best risk-reducer available. Entrepreneurs are risk-takers, but not needless risk-takers. Marketing research cannot guarantee success, but it can cut the chances of failure.

Consider this anecdote. A NASA scientist is also a baseball enthusiast. He has transferred some of his considerable scientific prowess to his baseball interest by developing a computer simulation technique that he feels could improve hitters' abilities to cope with pitchers. His invention enables him to simulate via computer the trajectory of an individual pitcher's fastballs, curveballs, etc. A hitter using the simulation could compress years of experience of hitting against a specific pitcher into a very brief period. The NASA scientist has also written and published a novel concerning the imaginary use of his technique. Neither the novel nor the technique has been well received. The scientist reportedly still believes that his computer-simulation method will catch on someday and that his situation is analogous to IBM trying without much success to sell computers for business use in the 1950s.

We have cited this example because it illustrates that even a little exploratory marketing research can be rewarding far beyond its costs. Exactly why the NASA scientist's invention has not been accepted, we do not know. But we do feel that unbiased trials by discussions with only a few major-league baseball players and/or clubs would in all likelihood have uncovered important objections to the computer-simulation hitting technique. The invention could then have been modified or aborted or marketed differently, depending on what was discovered. Enamoredness with a product, process, or technology is, from a business standpoint, dangerous. It leads the captivated individual or company to evaluate the potential for market success against personal criteria instead of against the criteria of the people whose opinions count most—the intended buyers.

Techniques for Engendering and Evaluating New Product Ideas

The oldest and most familiar method for stimulating creativity is brainstorming, which has been used in a diversity of endeavors by professionals and laymen alike. But there are other techniques as well. To what extent are these used by successful high-tech firms to come up with and analyze ideas for new products? Just how helpful has experience shown them to be?

The most popular creative technique for prompting new high-tech product ideas remains *brainstorming*. About 80 percent of the high-tech companies we asked have employed it and, of these, almost all of them have found it helpful.

Brainstorming designates a fast-paced group discussion on a predetermined problem or issue. For an hour or two, participants are encouraged to suspend their sense of logic. They are asked to let their minds wander and to suggest possible solutions, no matter how preposterous or absurd they may seem. Group members are urged not to make judgments about or to ridicule input from any group member.

We found that the next most widely used creative technique in experienced high-tech companies is the *focus group* approach. Nearly three-fifths of our firms have used this method and 80-plus percent of them have felt it to be helpful.

Focus group research involves selecting a small number of individuals, usually eight to twelve, that are felt to typify the overall characteristics desired. They might, for example, broadly parallel the target market or audience or they might be consumer or industrial innovators and opinion leaders. Then, the selected people are brought together to discuss virtually anything that the company needs to know to do a successful job of developing and marketing its offerings. Focus groups are not to be confused with the brainstorming technique; there are critical differences between the two. The word *focus* correctly implies that the group session is carefully directed. This direction or focus is provided by a skilled moderator, who cautiously but subtly moves the ostensibly free flow of discussion toward specific objectives which are not made explicit to participants. The focus group members are told the general topic of discussion but not the particulars.

In consumer-goods marketing, focus groups are scheduled at a variety of times and places. Industrial marketing (business-to-business) focus groups require more ingenuity in scheduling; potential buyers tend to be busier and more geographically dispersed. Trade shows, expositions, and conventions are good places to hold industrial focus groups. Teleconferencing is a possibility, but is advisable only as a last resort. It detracts from the group dynamics of a face-to-face discussion.

Focus group prospects are normally provided an inducement to participate—a monetary incentive or a gift. It is often felt that it may be difficult to entice people and professionals to serve in focus groups. Experience reveals that many of them are surprisingly willing to serve. One seven-member focus group included presidents and senior vice-presidents of large steamship companies. Similar results have been achieved with physicians, dentists, and other professionals.

Focus group sessions can be expected to last one to two hours. Anything shorter often does not allow group dynamics to take full effect and anything longer is likely to create wear-out. Sessions need to be taped for later review, either audio only or preferably both audio and video.

Easily accommodated with the focus group is the creative technique known as *synectics.* In an effort to provide a variety of creative perspectives, individuals with diverse educational and occupational backgrounds are selected for group discussion. The group leader is the only participant who knows the specific problem under consideration. He or she begins the session by asking the group to consider a broad topic that encompasses the problem but does not divulge it. For instance: "Why do surveys find that some occupations are so much more stressful than others?" As the discussion moves along, the group leader gradually directs the dialogue away from a general consideration of the elements that make for job-related stress to the hidden agenda, which is to answer the question, "What more can we at Hudson Computers, Inc., do to apply our technological know-how to make secretarial work easier and less monotonous?" (Surveys do, in fact, reveal the secretarial occupation to be one of the most stressful.) It is hoped that the stress elements identified by the general discussion will aid in triggering ideas to solve the specific problem.

About half of the experienced high-tech companies have used the creative technique known as *attribute listing* and most of them have found it helpful. Attribute listing has much to recommend it as a stimulus to creative thought for the high-tech firm; it facilitates thinking about how to apply existing technology better to solve a problem or to come up with alternative and superior technology for meeting a need or want.

To illustrate how attribute listing works, take the case of the now outmoded equipment that not long ago was used by clothing manufacturers to cut various sizes and styles from bolts of cloth for later assembly into finished garments. The major attribute of the old equipment would have been listed as shears or shearing capability. Companies such as Richmond Brothers now use laser echnology to do the cutting from the bolts of cloth in a fraction of the time. Once the attributes of anything are listed, the creative question becomes: What can be done to (or for) these attributes to change them for the better or, as in the case of the clothing shears, to obsolete them entirely?

Another creative method that has been tried by well over 40 percent of experienced high-tech companies is *forced relationships.* Of this 40 percent, about three-fifths of them have had favorable results with it in generating ideas for new products.

The technique requires an individual or group to list items or objects and then to consider the possible utility of the objects when combined in innovative ways. Word processing equipment resulted from some enterprising person picturing in his or her mind the combination of typewriter and computer.

The best-kept secret for prompting technological creativity is the method known as *morphological analysis.* Only a small number of proven high-tech firms have used this technique, about one out of five, but those that almost unanimously endorse its merits for stimulating ideas for new products. In fact, among high-tech users, it is the most valued creative technique.

Morphology, in the context of morphological analysis, refers to the form and structure of any given problem. The technique requires that a problem first be stated in general terms. Subsequently, it is dissected and its underlying make-up (its parts or dimensions) is identified. How each part of the problem might be (or has been) solved is profiled. Finally, these solutions are considered in combination.

Consider, for example, these underlying issues to a larger problem: the need for technologically complex and expensive equipment; the need to install the equipment and take precautions against its theft; the need to train for and pass a test before being licensed to use the equipment; and the traveler's fear of being stranded or having to walk or hitchhike for help. The larger problem is "How to assist the driver whose car breaks down on a trip?" The solution described by the foregoing issues is some type of communications equipment. Originally, the answer was impractical for the average individual because it required expensive equipment (as well as training, testing, and licensing) like that used on taxicabs and police cars. The innovative answer was the CB radio. Still, all the problems were not solved satisfactorily by the first round of innovation—problems such as the need to install the equipment (the radio and the aerial) on the car and the need to guard against theft. As a result, further innovation was forthcoming in the form of the hand-held CB radio that unobtrusively fits under the automobile seat. It is convenient, easy to use, and because of its portability, quickly protected against theft.

Another creative technique is the *nominal group*. Like morphological analysis, it has had only moderate usage in mature high-tech companies, but it receives strong endorsement from its users as a tool for triggering new product ideas.

The nominal group technique is a highly structured approach that has the effect of tempering the inordinate influence one or more individuals could have on the group were a less structured and more freewheeling group discussion to be used. Ordinarily, some subordinates might be overly supportive of the ideas advanced by their administrative superior. Or, subordinates could be inhibited by the boss' presence. Here is how the nominal group works.

Six to ten participants are seated at a table. A problem is presented by the group leader, but not in detail. Without any consultation or collaboration, the participants write down their thoughts about possible solutions to the problem. In round-robin fashion, these thoughts are presented for all to see, one per person per round, on a blackboard, an overhead projector screen, or a large piece of paper. While these presentations are taking place, no group discussion is allowed, but the person in front of the group is permitted to explain briefly what he or she has written down. Once all ideas have been presented, a very structured group discussion occurs, with each participant getting equal time. Ultimately, the ideas are evaluated by a rating or ranking procedure, whereby the participants vote silently, confidentially, and in writing. The pooled rating or ranking numbers indicate the group's decision about the relative merits of the ideas.

Additional techniques used by high-technology companies for prompting ideas for new products fall within the domain of day-to-day environmental scanning and marketing intelligence gathering. Widely used are: monitoring, indexation, and evaluation of ideas from relevant technical literature or other sources; scanning and assessment of environmental trends in such areas as competition, social-cultural evolution, and political and macroeconomic developments; and survey research to obtain the thinking and perspectives of customers, technical experts, and other potentially helpful individuals and groups.

Premier high-tech companies incessantly seek ideas and information from a variety of sources. The available creative techniques are not treated as though they were mutually exclusive; the use of one does not preclude the use of another. As experience in high-tech companies has shown, a number of these creative techniques are valuable for triggering ideas for the new products and processes upon which the high-tech firm depends.

Who to Ask

Regardless of the creative technique used—brainstorming, focus group, or whatever—who do high-tech firms need to ask for opinions about possible new products? Conceivably, individuals with a variety of traits, backgrounds, and occupations could prove helpful. In fact, high-tech companies typically do look to several sources of opinion and creativity to provide them with a stream of new product ideas. They probe the thinking of their customers; their own sales representatives, marketing executives, and R&D personnel; external consultants and experts in the scientific community; and others such as the company's top management and project managers.

For all high-technology companies, it is both prudent and intelligent to seek new product ideas from a diversity of sources. A focus group comprised of sales reps could prove useful by itself, but additional and different perspectives from a second focus group with R&D personnel might contribute synergies or reveal why ideas from the sales reps would not work in practice. The search for new and creative ideas is massive, systematic, and endless in any high-tech company that plans to stay around as a competitor in the true sense of the word.

Even though good marketable ideas and the evaluation of those ideas may be obtained from numerous places, we believe strongly that the best single source is the consumer or industrial innovator. Market research into product conceptualization and development without heavy focus on innovative buyers is woefully lacking in substance, especially in high-tech markets. As demonstrated time and again by diffusion research, the theory of the vital influence of the buyer innovators is well supported by empirical fact. The initial 10 to 15 percent of the customers who adopt a technological innovation as their own can, through their personal example and influence, determine how many buyers ultimately follow suit and how quickly the diffusion process unfolds.

Whether or not there is a disproportionate number of innovator firms or individuals among potential high-tech product purchasers has not been adequately researched. Much additional research is needed to determine if this conjecture might be correct, especially in the business marketplace.

What has been the experience in proven high-technology companies with the use of buyer innovators? Have these companies attempted to identify innovative consumers or industrial users to ask about or test new products? If so, has it improved their "hit rate" in ascertaining how new products might or will fare in the marketplace?

Not surprisingly, because diffusion theory and research apply precisely to technological innovations rather than to annual model changes and other so-called new products, most high-tech companies are well aware of the buyer innovation concept and its value to them. This holds true across a diversity of industries, including those that sell to the ultimate consumer and those that market to other businesses or institutions. About three-fifths of proven high-tech marketers have tried to identify innovators for the purpose of testing new product ideas. Of these, nearly 80 percent report that their "hit rate" has been improved accordingly. For instance, a pharmaceutical executive said, "We use key opinion leaders, usually research-oriented physicians." An industrial marketer indicated that, "Leading-edge customers (*Fortune* 500) are great sources" of new product ideas. One executive lauded the innovativor technique, but added a caveat: "This method is most productive but is expensive and time-consuming."

By some standards of judgment, the identification and use of buyer innovators may seem expensive and time-consuming. Yet, in a relative sense, compared to the money and time typically lost on a product failure, the innovator method is usually a bargain. It is a known risk-reducer. If a new product idea does not "fly" well with innovators initially, this bit of information raises a red flag about the future of the idea if developed and commercialized. What reason is there to believe that a high-tech product will be accepted and succeed later if it is generally rejected at the initial stage by the most change-seeking, innovative, and influential of possible buyers? The use of innovators to screen product ideas is particularly valuable in warning the high-tech company that it may very well have a "losing" new product idea. The time and money expended in mitigating the considerable downside risks of new product failure intrinsic in high-tech industries are small prices to pay. The firm that goes the extra step by seeking information from innovators is on the right track. It will not guarantee success, but it will reduce uncertainties.

The process of identifying innovative buyers who would make knowledgeable candidates for group discussions about new product ideas is usually not that arduous. In business-to-business market situations, buyers are often not numerous and, thus, the job of discerning and eliciting comment from innovative industrial buyers is direct. As the aforementioned executive in a high-tech company pointed out, *leading-edge* customers (*Fortune* 500) are good sources. For consumer-goods firms that market to ultimate consumers rather than to other organizations, the task of identifying innovators normally becomes more complex; there are more potential buyers from whom innovators must be identified. Moreover, consumer markets are unlike industrial situations where it is fairly clear as to which buyer might qualify as an innovator, for example, a person with the title of physician or mechanical engineer or head of the accounting department. The person's education and position designate his or her specialty. In consumer markets, a person has no specific title to hint at which general areas he or she may be considered to be an innovative buyer. However, even here, the task of identifying innovators is not all that difficult. Whether the high-tech company is concerned with industrial buyers or ultimate consumers, *the job of finding innovators is basically one of looking for buyers who have been early purchasers of products of the same general genre in which the proposed new product falls.* If the product is personal in-home computers, look for electronics buffs; if the product is robotics, seek out manufacturing engineers who work for companies that historically have been in the vanguard in installing new methods of productivity; or if the process is biotechnology, look for scientists or physicians, as the case may be, who have been demonstrably receptive to change and willing to promote it.

Purchased mailing lists can provide leads to innovator identification. Subscription lists of technical publications and membership rolls of specialized and often esoteric scientifically oriented associations and clubs usually include a large number of innovators. Similarly, customer lists of specialty firms likewise provide leads (for example, Radio Shack). The Touch Tone Telephone was test marketed to innovators in the Chicago area whose phone company customer records revealed to have previously adopted other phone innovations quickly. A computer-based information storage and retrieval system in education—the U.S. Office of Education's Educational Resources Information Center—was used in its early days by one of us to identify innovative secondary and elementary educators. As predicted, these teachers had a history of being innovators across a broad range of instructional aids. What is more, they were, as expected, opinion leaders to whom other teachers looked for advice. They also had demographic and attitudinal traits in common.

Referrals are yet another way for the high-tech company to go about finding innovators. Certainly in business-to-business marketing, sales representatives should readily be able to pick out innovative industrial buyers and companies. In addition, the self-report technique, in which a person is asked a series of questions about his or her innovativeness and opinion leadership traits, fortunately has been demonstrated to be a valid method for screening innovators from noninnovators with regard to a general category of products. This self-report approach is particularly useful in identifying innovators in consumer-goods industries.

So, while identifying innovators and getting them to take part in a group discussion of some type is time-consuming and sometimes necessitates considerable ingenuity, the effort is usually worth it. Their opinions and ideas are the best form of insurance available against new product failure in the market.

Occasionally, a marketer of high technology will say that innovators are not appropriate opinion-providers for his or her firm. One executive told us: "Innovators may not be typical of overall market results in small-volume specialty sales." We could not agree more. By definition, innovators are never typical of the individuals in the total market. No, instead, innovators are far more influential and, through trying new products early and by word and example, they determine whether and how rapidly a new product moves through its diffusion process. Without the cooperation of innovators, a new product never gets off the ground. And without favorable word-of-mouth or similar personal reports from innovators who have tried a new product, it rarely survives. That, in capsule, is why the opinions of innovative buyers need to be weighed so heavily in creating and evaluating ideas for new high-tech products.

Estimating Market Demand

Once product ideas have been elicited and evaluated, some of them are likely to be considered worthy of retention and further exploration. At least that is the hope. From those ideas that pass the initial screening, management must somehow make rudimentary forecasts of market demand. A number of techniques are available for gathering data on which these forecasts can be made. These methods are not mutually exclusive; the use of one does not preclude the use of another. On the contrary, savvy management obtains data for making market demand forecasts in a variety of ways so that cross-checks are provided and possible product deficiencies are discovered and hopefully overcome sooner rather than later.

After a product idea has made it through the initial screening stage, how do proven high-tech companies go about estimating market demand, which is another way of asking how they forecast the degree to which a product will succeed? To answer this question, the new product development process needs to be divided into early stages and later stages. In the initial stages immediately following idea screening, certain techniques are appropriate, whereas downstream toward commercialization, different success-forecasting methods are more helpful.

Early on, the high-tech company is concerned with whether the basic concept underlying the product or process is one that will fulfill a significant market need or want better than any competitor product or technology. The more revolutionary the concept, the more difficult this issue is to address. In the case where a concept will, in effect, create a new market, the process boils down to what one executive in a high-tech company called "sophisticated wild guesses." In other words, even though the best of market research techniques are used to assess a product's potential and reduce the risks of market failure, no one can say with any surety that the product will create a certain market demand in a specific time frame. State-of-the-art market research techniques lend sophistication, but there is a strong element of guessing in the ultimate decision.

It is in this kind of uncertainty that the true entrepreneur and his or her sense for the market surfaces. A decade ago, in one company a nine-person top management policy board unanimously counseled the firm's chairman and chief executive officer not to reintroduce an old product with a new technology. Their advice was based on long experience and extensive market analysis by corporate staff. The chief executive ignored this advice and directed R&D and marketing to proceed full steam ahead. The technology behind the product has made it today's market leader and has largely obsoleted the technology of ten years ago. Some decisions are so entrepreneurial that they almost defy traditional business analysis and thus must be made on feel for the market and high tolerance for risk.

Fortunately, only a very few high-tech products are so revolutionary that sophisticated wild guesses must be relied on. Most high-tech products are but new and innovative applications of existing or modified technology, hence there is at least some track record for the technology from which management can make accurate inferences.

Like any other method of business analysis, marketing research is meant to be an aid to decision making, not a substitute for it. Information and findings from marketing research can help management make better decisions, but they cannot guarantee them. And this is true many times over for high-tech firms. If business research techniques and computers could make decisions for management, there would be no need for human executives. What is more, it is doubtful that computers could ever be instilled with that almost mysterious quality of entrepreneurial flair—the "animal spirit" referred to by John Maynard Keynes.

Concept testing and *product prototype tests* are the major marketing research techniques used by high-tech companies in the early stages of the product development process. Once a potential product has passed the idea screening stage, virtually all high-tech firms turn to concept and product prototype testing. In fact, high-tech executives have found them to be the most helpful techniques available for predicting whether or not a product can be a market success.

In concept testing, the basic idea or theme of the product is developed and evaluated. No physical product exists yet; the concern at this juncture is solely the concept. Questions are explored about what needs and desires the product might fulfill and how it could be positioned and promoted vis-à-vis competitive means to meet the same needs and preferences. Without a sound concept of what a product is all about and why it will be valued in the marketplace over competitive offerings, there is scant chance tht success is attainable.

Once the right concept is agreed on, a physical product is developed in the form of a prototype or mockup. There may well be several versions. These prototypes are tested on individuals and groups that are thought to be capable of providing insightful and sound opinions about the product and how to improve on its merits.

Focus groups are the preferred means among the high-tech companies in our research for conducting both concept and product prototype tests. This technique is considered to be the best group evaluation method available, although perhaps 20 percent of high-tech firms have had success with the *nominal group technique.* On balance, however, focus groups are preferred by wide margin. The group dynamics encouraged by the focus group format seem to work to promote more useful information about concepts and products, in comparison to the more passive nominal group method of eliciting information.

Among successful high-tech companies, the experience has been that members of their own sales forces make the best in-house (employee) participants in focus group discussions about product concepts and prototypes. Although other individuals in a company's employ can provide keen insights, sales reps are the employees most frequently asked for their opinions. Sales reps' close proximity to the market on a daily basis no doubt accounts for the high value management places on their views. Highly sought after nonemployee focus group participants for concept and product prototype testing are, not surprisingly, buyer innovators. Because of the role and importance of innovative customers in influencing other potential buyers and in trend-setting by example, the value that high-tech companies place on their opinions is well-founded.

Consumer or industrial panels (depending on whether the product is a consumer good or industrial good) have been used by about half of proven high-tech marketers to test new products. Nearly three-fifths of them have found the panel technique helpful in making better product development decisions. In this method, the new product is not actually commercialized. But it is placed with ultimate consumers or industrial buyers like those in the market of concern. Panel members try the product for a while, in their daily lives for ultimate consumers and at work for industrial buyers, and then their comments and opinions are solicited and used by the high-tech company to make further decisions about product development and the advisability of commercialization.

Further downstream in the product development process, closer to commercialization, there are two marketing research techniques that are popular with high-technology companies in obtaining information on which to forecast a new product's potential in the marketplace. One of them is exhibiting at *trade shows* and the other is *test marketing.*

Trade shows are widely used by high-tech companies, especially by the industrial marketers among them. These exhibitions are a valuable forum both for showcasing a new product and for eliciting comments from possible buyers before a go/no-go decision about commercializing the product is finalized.

Test marketing is a method for forecasting a product's future and for experimenting with alternative marketing strategies. In this technique, a new product is actually commercialized on a limited scale to determine how it is received by buyers who are felt to be representative of those in the larger market. Not all high-tech companies have the option of this business version of a dress rehearsal because of the nature of the product or technology at hand. It is not feasible to test market many types of custom-made industrial products with extremely large unit values and limited demands. Sometimes even when test marketing is feasible, it is ill-advised from a competitive standpoint. In high technology, there is a premium on innovation and ingenuity. Often a company would like to test market, but goes directly to full-scale commercialization in order to be the market leader and not provide the competition with product development catch-up time. Of the successful high-tech companies that can and do use test marketing, which is about half of them, the vast majority feel that it assists them in making sounder judgments about a new product's market potential.

A forecasting method especially suited to predicting market demand for high-technology products is the *Delphi Technique,* which was developed by the Rand Corporation, a think tank of high repute. Delphi, after the prophetic Greek god or oracle, can be used at any stage of product development, from idea generation through commercialization. This method seems almost made for the special needs of the high-tech company because it can be used when there is a dearth of information on which to formulate estimates. Basically, when adapted to market forecasting, the Delphi Technique is a formalized jury of expert opinion in which a small number of individuals who are very knowledgeable about a particular market participate in a multiple-step estimation exercise. They initiate the sequence by providing their written estimates of market demand in a specified period of time. No discussion among the experts is allowed and individual estimates are contributed anonymously. After each of several successive rounds, the experts' estimates are given to every member in writing and in summary form. Following a few rounds, the predictions tend to converge toward what is hoped will be an accurate estimate based on the collective wisdom of the experts.

The Delphi Technique is similar conceptually to how the pari-mutuel system works at a thoroughbred racetrack. By betting, the race crowd affects the odds shown on the odds board for each horse. Eventually, the bettors converge on the horse that represents the crowd's best estimate of which animal will win the race. Even in this situation, where there is far more collective ignorance than wisdom, most bettors are not experts, the favorite wins about a third of the time.

Surprisingly, the Delphi Technique is largely unknown to the executives in high-tech companies whose positions require them to forecast market demand for new products. In the 20 percent or so of proven high-tech companies that have used the Delphi Technique to forecast a product's destiny, the experience has been mixed. There is about a fifty-fifty split between those companies that have found it to be a valuable forecasting tool and those that have not.

Our judgment is that the Delphi Technique, if properly executed, has much to recommend it to the marketer of high technology. Unlike more quantitative forecasting methods, which usually predict the future from past data, Delphi requires little historical data. The lack of meaningful data often accurately characterizes the situation whenever a new high-tech product is involved. Moreover, the technique formalizes the thinking of executives by requiring them to commit their thoughts to paper in the form of a number estimate. What is an excellent, good, moderate, fair, or poor market demand is subject to various interpretations; but a demand of ten thousand units of sales in the first year of commercialization is not. The Delphi Technique also enables an executive to see what his or her colleagues are thinking about a new product's potential. The noncollaboration and anonymity provisions keep a dominant or persuasive individual from inordinately affecting the estimates.

Irony of Ironies

By its very nature, high technology rests solidly on the foundation of science and mathematics and the precision and exactness they connote. Yet, our findings are quite clear that the marketing research techniques most appropriate to supply-side high-technology markets fall mainly in the domain of qualitative rather than quantitative marketing research. (Of course, as the market becomes more fully developed and there are performance records, more traditional marketing research techniques can be effectively employed in order to determine the type of demand a state-of-the-art improvement might stimulate.)

This result is ironical but not really startling or even unexpected. Many of the more mathematically based methods of marketing research used in mature industries, as well as the popular econometric models for forecasting, require plenty of data—often data obtained from a random sample of people so that statistical inferences can be made. This requirement is difficult to meet in the supply-side stage in high-tech industries for two reasons. First, whenever markets are being created or obsoleted rapidly by significant product breakthroughs, as they are in high technology, not much is available in the way of reliable historical data. An executive in the small computer industry remarked that his firm does not purchase much in the way of forecasts, because projections are almost immediately obsolete due to rapidly changing market conditions.[2] Second, data obtained from prospective buyers via traditional data-collection techniques (telephone interviews, mail surveys, etc.) are of dubious value for answering questions about products based on new technologies. The product-education opportunities and in-depth probing afforded by qualitative research, such as group discussions, mitigate these kinds of problems. For instance, it might be more desirable to have opinions from ten buyer-innovators gathered in a two-hour focus group than to have opinions from a random sample of one thousand people, most of whom can only vaguely conceive of a new product, its technological benefits, and how it will fit into their life styles. Or, the Delphi Technique, as used by a jury of six experts, is likely to give more accurate estimates of how successful a significant new product may be in the marketplace next year than are the predictions of a sophisticated econometric model that must be fed a plethora of historical or cross-sectional data. But this subjective approach to marketing research in high-tech industries is not really a problem at all. As pointed out by the marketing director at Apple Computer, strategic planning does not require precise forecasts of marketing demand; precision is not needed when a market's growth rate is exploding. From a practical managerial standpoint, it matters little or at all whether a market is growing at 60 percent or 80 percent annually.[3]

As any high-tech market matures, there are increased possibilities for the use of some of the exciting techniques being employed in consumer marketing. The multivariate techniques that have become a vital part of the consumer behavior literature and practical research are especially valuable in revealing subtle differences that might be overlooked by qualitative marketing research methods. In addition, as high-tech markets evolve from supply-side to demand-side conditions, it is permissible to use the random sampling and inferential statistics that enable one to make more valid generalizations than are permitted with qualitative approaches.

Notes

1. "Listening to the Voice of the Marketplace," *Business Week,* February 21, 1983, pp. 90–91.
2. David Stipp, "Firms Researching the Market in Small Computers Grow Fast," *Wall Street Journal,* May 27, 1983, p. 21.
3. Ibid.

5 Managing High-Tech Products

No one is more blind than he who does not wish to see.

Volatile market conditions in technology-oriented industries make product planning and management the most challenging element in the high-tech firm's marketing mix. As experience after experience corroborates, entire industries can be revolutionized or swept aside virtually overnight—obsoleted and destroyed by a single technological breakthrough.

Often, high-tech firms' fortunes ride on how well they innovate vis-à-vis the competition; new products are their economic lifeblood or their offensive thrust. Yet, any firm can look good as a front runner with a useful technological innovation to offer. What separates the bona fide strategic marketing managers from the pretenders is how well a firm suffers adversity and responds to superior technologies innovated and commercialized by competitors.

The preponderance of research on technological innovations has dwelled on the experiences of those companies which have been pioneers in the research, development, and commercialization of technologies. Yet, much is to be learned from the experiences, particularly the mistakes, of companies in the mature industries which have been imperiled by new technologies.

What we think is an extremely insight-providing study, by Professors Arnold Cooper and Dan Schendel of Purdue University, looked at case histories of 22 firms in seven different industries that were partially or wholly supplanted by innovations.[1] They studied, for example, steam locomotives versus the diesel-electric, vacuum tubes versus the transistor, and aircraft propellers versus jet engines.

Based on Cooper and Schendel's in-depth examination of the process wherein a new technology comes onto the market and begins to substitute itself for the old technology, several generalizations were made. They included:

> The majority of the time, new technologies are initially commercialized by companies *outside* the threatened industry. Start-up firms are especially likely to innovate new technology whenever capital requirements are not huge.

Frequently, because innovations early on are typically crude and expensive, pessimistic prognoses for the new technologies in the marketplace are common. (One early and supposedly authoritative forecast for computers predicted a market potential of 50 to 100 units.)

Innovations often create new markets that are not open to the replaced technology; vacuum tubes were incompatible with most of the equipment made feasible by the discovery of the transistor.

The new technology frequently follows a roll-out or domino market penetration process, whereby it sequentially seizes market segments from the threatened industry.

Sales over time do not always follow the classic s-shaped growth curve (see figure 5–1). This observation holds for both the new and old technologies.

After the new technology's introduction, sales of the older technology do not invariably decline right away. On the contrary, revenues often actually increase for the older technology. Once revenues of the displaced technology do begin to decrease, it takes anywhere from five to fourteen years for the sales of the new technology to surpass sales of the old technology.

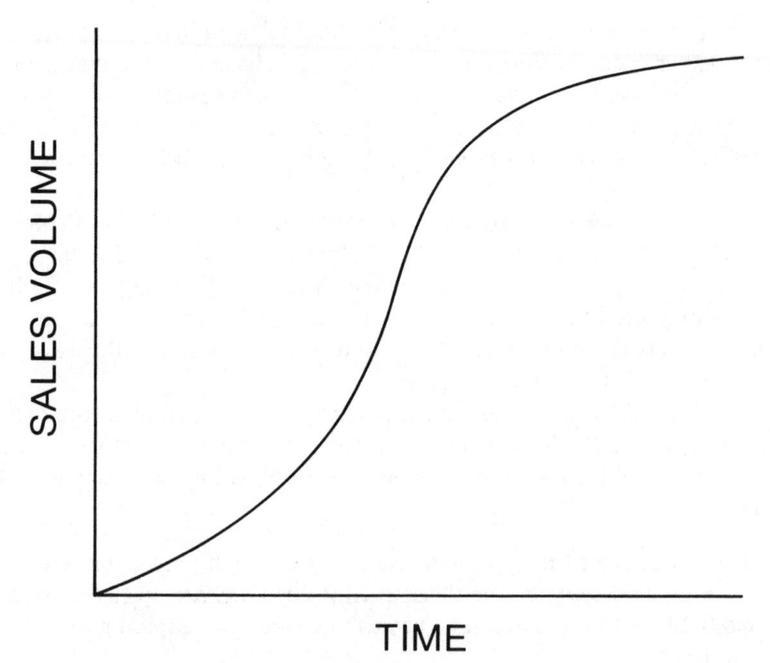

Figure 5–1. S-Shaped Sales Curve

Only 5 of the 22 threatened companies studied by Cooper and Schendel elected to forego participation in the new technologies ominously confronting them. Of the 17 companies that did elect to compete in innovative technology, 15 made major efforts. Some of these were early entrants into the new technology and some were laggards who came in years after the innovation was commercialized. Significantly, we think, acquisition was not a popular method used to establish a firm's presence in a new technology.

Without exception across the 22 endangered companies, the old technology was perfected to its fullest extent *after* the new and superior technology was brought on the market as competition. The smallest and most dependable vacuum tube came after, not before, the transistor's emergence in the market. In no instance did a threatened company directly abandon the old technology in favor of an undivided effort to develop and market the superior technology. In fact, most of them did exactly the opposite: They heavily committed resources to improving the obsoleted technology, even after company revenues had fallen off due to the competitive inroads made by the new technology. The 22 companies generally hedged their bets by ambivalently dividing their resources between the old and new technologies, so as to participate in both. In comparison, the firms that pioneered the new and superior technology rarely tried to enter the older and inferior technology.

What resulted? How did the endangered companies fare? Not well at all. Most of the 22 firms were unsuccessful in obtaining strong positions in the innovative technologies that threatened them and few achieved market-share leadership positions.

Richard N. Foster directs the consulting firm of McKinsey and Company's involvement in technology management. By his estimate, the U.S. companies that for years have remained on top as technological leaders have been the innovators of new technologies in their respective industries about 30 percent of the time—a remarkable record. In spite of the fact that long-term technological forecasting is difficult at best, Hewlett-Packard, IBM, Dow Chemical, Eli Lilly, and those few others like them, have not yielded their leadership positions. In Foster's view, their continuing success is attributable to their recognition of what McKinsey and Company sees as the five realities of technically based economic competition.

1. Technological leaders that are successful in staying on top never lose sight of the reality that all products and processes have performance limits. As a technology approaches these limits, it becomes more expensive to make subsequent performance improvements. As a result, the leader looks for new and better technology.

2. Technological leaders do not underestimate their competition, regardless of how small that competition may be at the present time. A

nearly bankrupt Boeing introduced the jet plane and promptly overcame the industry leaders, Lockheed and McDonnell Douglas Corporations.

3. Technological leadership requires that a firm aggressively pursue and invest in potentially superior new technologies.

4. This aggressiveness should come early on because the process by which one technology substitutes for another begins slowly and then explodes, usually in an unpredictable way.

5. High-tech firms that maintain leadership positions have close collaboration between the chief executive officer and the chief technical officer. This arrangement helps to assure that technical programs are responsive to business strategies and vice versa. Yet, only about 20 percent of the CEOs in U.S. companies today have their top technical executive as a member of the top management circle.[2]

Leslie Cook is the president of a consulting firm which specializes in R&D programs and management; he is retired from his former position as manager of research program planning for the Exxon Corporate Research Laboratory. In Cook's experience, notwithstanding the more acceptable business and technical arguments usually advanced as rationale, the overriding factor in whether or not a company innovates a new product or process is the CEO's perceived personal risk.[3] But the risk of concern *is not* that the innovation will fail technically or financially. Instead, the CEO is usually motivated by and acts on the way he or she answers this question: "Does the incremental risk to him personally of trying for the innovation increase—or reduce—his overall risk situation as he presently perceives it?" A perceived increase in incremental personal risk will make the CEO averse to the innovation, no matter how large or how sure the monetary return on it is projected to be. By contrast, a conclusion that an innovation will reduce his overall personal risk makes the CEO interested in moving ahead with the innovation, irrespective of the technical and financial uncertainties associated with it.

For R&D executives, understanding the CEO's real underlying motivation for saying yes or no to an innovation is essential. In Cook's long experience, understanding is essential if the CEO is to be persuaded of the innovation's merits and sold that going forward with it reduces the company's (and his) overall risk. Cook has observed that nothing confuses an R&D manager as much as for the company's chief executive officer to reject a commercially promising innovation—often an almost sure financial success—after years of research and development effort. In extreme cases, R&D's success has resulted in the breaking up of the R&D group because, in the words of one executive, "What is the use of an organization that comes up with successes we do not want?" Once the company's existing technology is in jeopardy from competition, the CEO becomes more innovative.

In Bell Labs, the transistor was pursued because of the risk that vacuum tube switching was thought to constitute for AT&T's long lines investment. And, the transistor was pushed before there was any certainty that it could be achieved technologically.

Cook can catalogue numerous examples of where a company held back a new technology because, if commercialized, it might fail or would obsolete the firm's existing products and thereby increase the CEO's overall risk. But, whenever competition seemed to be on the verge of commercializing a similar innovation, the endangered firm moved forward because now the old technology and the CEO's personal situation were at major risk.

The trouble with a CEO waiting to focus on new technology until the company's existing technology is competitively disadvantaged is that, by then, the competition can have built up insurmountable market leadership. If the need arises, a General Electric may have the know-how and resources to engage in a crash program that enables it to catch competition or to acquire a firm already in the desired industry. But the GEs of the world of high technology are the exception; and even General Electric, with all its expertise and financial reservoir, found that it could not become a potent force in computers—it started too late to catch the leaders. Most firms that lose market leadership early on either remain as market followers or withdraw from the market entirely.

A saying goes, "No one is more blind than he who does not wish to see." Success often breeds self-induced blindness through complacency and an affinity for the things that brought about prosperity. It is easy to describe the portrait of a heretofore successful CEO in a high-tech firm who is about to provide additional evidence of the proverb's validity. He has predictable ways: The CEO is enamored with a technology and, quite possibly, was in on the ground floor of its development and commercialization. When confronted with threatening new technology, he strives mightily to recoup or enhance his company's initial investment in the existing technology before sloughing it off. The CEO hedges his bets by simultaneously going all out to improve the old technology and to sell it aggressively—while directing R&D to begin work on the newer technology "just in case." His resources are thus divided between competing interests. The CEO's strategic compass is guided by the experience curve; the economic phenomenon whereby the firm's real costs (inflation-adjusted) of manufacture and marketing fall with every doubling of units produced. The CEO's company is a market-share leader and therefore has a significant price advantage over all or much of the competition due to efficiency and lower costs yielded by the experience effect. The CEO does not intend to give up this advantage without a fight that he cannot win.

All of the CEO's reasoning and rationale dwells on a past that is irretrievably gone and rendered irrelevant to future competitiveness by superior technology. The dispassionate free market will reward better products and processes; it does not take into account, nor should it, one firm's sunk costs and capital investments, experience effects, and management's affection for and long association with "the technology that made us what we are today."

Regardless of the degree of sophistication of the business and technical sides of a high-tech company, whether the company leads and sustains that leadership hinges ultimately on the innovativeness and creative destructiveness of its CEO. The best R&D personnel and professional managers that money can hire will not save a market leader whose top executive is wedded to the past and thus manages from a short-run, risk-avoidance strategic frame of reference.

The Eli Lillys, Hewlett-Packards, IBMs, and other members of the high-tech *crème de la crème* who have managed to sustain their success are fundamentally different philosophically. They are willing and able masters of creative destruction of their own capital. Their orientation is toward the market; and market realities rule with an iron hand over the experience curve. Efficiencies are not ignored by any means, just relegated to second place behind market effectiveness, where they belong. These companies matter-of-factly assume that their proven and profitable products and processes of the present are those most vulnerable to obsolescence. They act on the economic maxim that profitability invites both competition and the introduction of new, improved, or even revolutionary ways of doing things through technology.

But, as a practical matter, how does a company recognize that its technology, or better yet a competitor's, is in danger of obsolescence? Surely not by looking at economic signals. Experience teaches that it is frequently after the introduction of a new and improved technology that an older technology reaches its peak in terms of both technical performance and financial rewards. A moribund technology can appear deceivingly healthy, for a time. If a firm waits for economic trends to augur impending doom for its existing technology, the firm will be late and maybe too late to move into new technology. Competition will already have established a strong foothold in the new technology.

What, then, is the alternative to using economic signals as an early warning system? Based on empirical studies, several consulting firms—primarily McKinsey and Company and Arthur D. Little, Inc.—believe that the monitoring of relevant technology life cycles is the best way for a company to provide itself with ample warning that a product or process is imperiled. For example, look at figure 5–2. The horizontal axis designates units of effort that have been expended on a technology to date, such as the total R&D man-years or dollars spent. The vertical axis refers to the most

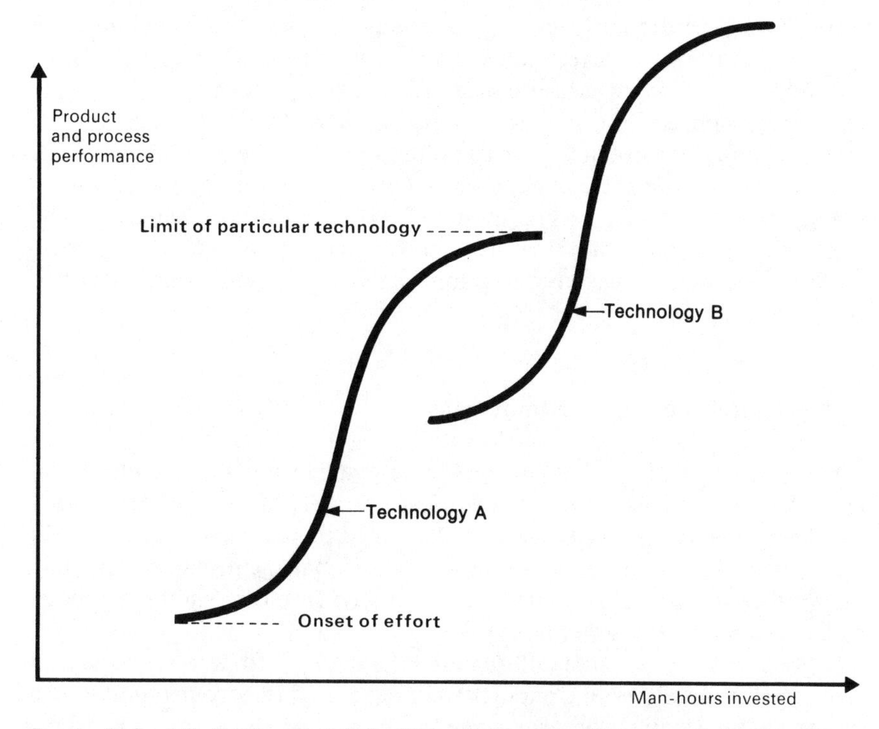

Source: Richard N. Foster, "A Call for Vision in Managing Technology," *McKinsey Quarterly* Summer 1982, McKinsey and Company, Inc. Originally appeared in *Business Week,* May 24, 1982. Reprinted by permission.

Figure 5-2. Technology Life Cycles

crucial measure of a product's performance; it might be melting temperatures for a plastic or bytes for computer-storage devices. Technology A starts out slowly, then because of intense and heavily financed R&D efforts performs markedly better, and finally plateaus as the technology reaches its performance limits. At Technology A's performance capacity, the new Technology B is already wending its way through its own s-curve, a curve that starts out at a performance level far above Technology A's origin. For a while the two technologies compete (their s-curves overlap), until the superior Technology B eventually wins out and captures the market.[4]

The high-tech company that is successful in staying on the cutting edge of its industry's technology must spend more time analyzing technology life cycles than on studying experience curves—much more. This is not to say that attention to efficiency and low-cost manufacturing and marketing is unimportant for the high-tech firm. Far from it. Rather, the point is quite simply that the changing marketplace and appropriate market strategies and tactics take precedence. It does the company little good to be the low-cost producer and seller of a technologically obsolete product or process.

What is the critical measure of a technology's performance? Is the old technology approaching its performance limits? Has R&D been able to make meaningful incremental progress in improving the technology? And, most salient, what is the comparative performance of the old and new technology or technologies on the all-important performance criterion? The willingness and ability of a firm's top management to ask and answer these kinds of questions objectively go a long way toward determining the company's destiny.

Technological Innovator or Imitator?

The high-tech company that manages to innovate technology usually has an exploitable advantage over its less creative rivals. The value of market leadership has been proven time and again. Yet, there is a profitable place in many markets for product imitators whose managements intentionally *never* innovate technology. Not every high-tech firm can, should, or needs to be an innovative market leader.

Much has been written and said about Japan's efforts to seize the lead from the United States in computer technology and to become number one in the global information-processing industry. For example, there is the well-known Japanese drive to develop leadership in so-called intelligent computers that can converse with human beings in nontechnical language, as well as in the technical idiom of a particular occupation. The Japanese are making a similarly ballyhooed all-out effort to gain worldwide superiority in the semiconductor industry. But what is sometimes overlooked in all the publicity about Japan's technological prowess is that the Japanese have historically been technological *imitators,* not innovators. As Herbert Schorr, vice-president of systems research at IBM, has remarked about the Japanese endeavor to develop human-conversant computers and a first-rate research capability: "They haven't done that before."[5] Much of the Japanese success until now has been in first copying others' technology and then in improving its application. The idea that the Japanese have been technological pioneers is widely accepted, but mistaken.

The lesson is: While constant striving to be an innovator is essential policy for a pace-setting high-tech market leader, technological innovativeness is not mandatory for all high-tech firms, especially for relatively smaller companies. A market follower can be a viable competitor in any market which allows for product differentiation—in other words, in any essentially noncommodity market. In these situations, a focused or niche market strategy is possible. The smaller firm succeeds by focusing or targeting its efforts on a market segment—a niche—from the total market. (In commodity markets which, by definition, do not permit significant product differentiation, market followers have little chance to be competitive; the cost/price advantage that accrues to leaders due to experience curve effects overwhelms low market share companies.)

An engineer from a prominent technological leader that manufactures and markets electric motors told us of his company's continual battle with technological imitators. By industry norms, his firm spends large sums of money on research and development, only to have the resulting technological improvements copied by foreign competitors. In a number of countries where patent and brand protection is weak or virtually nonexistent, competitors of technological innovators, in any industry, are often able to achieve, with impunity, blatant forms of copying. Take the case of Apple Computer lookalikes that are counterfeited and sold in several of these nations. The technology is not the same, but outwardly, the product closely resembles and is represented to the prospective buyer to be a genuine Apple Computer.

A vice-president of marketing and development in a U.S.-based high-tech firm characterizes his company's strategy as being "to develop and market products which either exist in the marketplace already or have clear market appeal because they are offshoots of existing products." He concludes, "so our hit rate on market success is high." The president of another high-tech company describes his corporation's market strategy very succinctly: "Watch the competition and clone its profitable products." For these firms, the product-imitator strategy has been exactly the right recipe.

Companies that are successful in being the innovators of technological breakthroughs in their respective industries even a third of the time are generally thought of as leaders. So, typically, the most innovative of high-tech firms will have occasion to play technological catch-up. We are of the opinion that, in such cases, these companies have been too inclined to catch up through internal R&D crash programs when external acquisition of the needed technology would better serve their purpose. Acquisition would be more expedient and, quite likely, less risky and not as costly. And, because small, entrepreneurial firms frequently are the originators of technological achievements, external acquisition is usually a feasible option for the large high-tech company that quickly wants to obtain a technological capability.

Acquisition also is often a better strategic choice than internal R&D for large companies that want to enter a high-tech industry for the first time. General Electric, for instance, might well have averted its ill fate in the computer industry had it acquired a "going concern" with the requisite technical and marketing know-how. It is not inconceivable that with GE's wherewithal and reputation it might have been able to acquire the then-fledgling but now prominent Digital Equipment Corporation.

Beware of Using the Product Life Cycle for Strategy Formulation

Proponents of the product life cycle concept say that, like living things, products proceed through predictable phases from the time they are born until the time they die. The familiar bell-shaped curve normally used to illustrate this process is shown in figure 5-3. One executive wrote of the practical managerial value of the product life cycle idea: "The product life cycle can be the key to successful and profitable product management, from the introduction of new products to profitable disposal of obsolescent products."[6] This sort of claim is fairly typical.

Yet, what is the empirical support for the bell-shaped product life cycle? A study at the Marketing Science Institute, near the Harvard Business School, looked at some one hundred product categories in consumer goods to determine whether they followed the introduction, growth, maturity, and decline stages, as depicted in the bell-shaped curve. The researchers provided damaging testimony about the accuracy and managerial applicability of the product life cycle and went on to say that the product life cycle may be more misleading than useful.[7]

Two executives at the well-known J. Walter Thompson Advertising Agency researched life cycle case histories in four diverse product classes. Their conclusions were likewise negative about product life cycle usefulness.

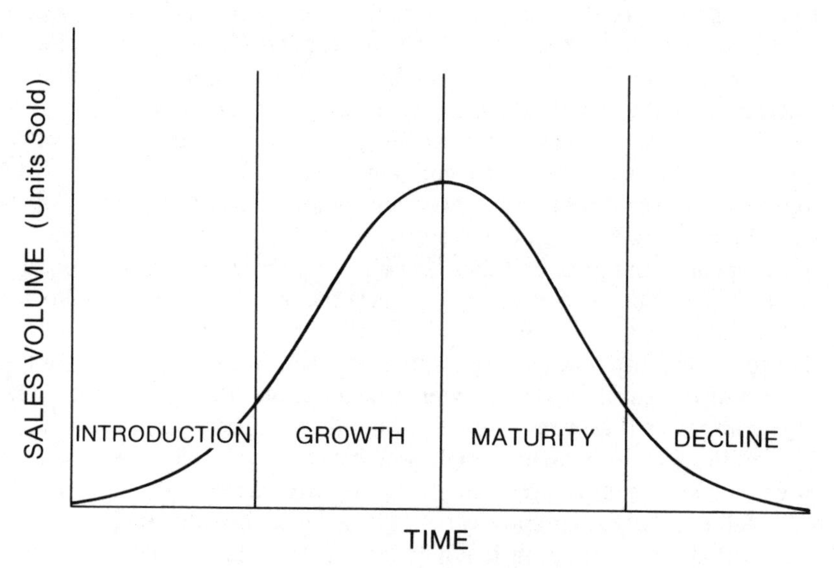

Figure 5-3. The Product Life Cycle

The J. Walter Thompson executives were of the opinion that no rules exist or can be found that indicate when a product is moving from one life cycle stage to another. What is more, it is difficult to project when a subsequent life cycle stage will appear, what its duration will be, and to what peak sales will go. It is not possible to evaluate accurately which life cycle phase a product is in because the major stages of introduction, growth, maturity, and decline do not neatly compartmentalize.[8]

Consider the divergence of opinion that has emerged about the semiconductor industry in the early- to mid-1980s. To some knowledgeable observers, the industry is maturing. The top official of the consulting firm of Arthur D. Little in San Francisco sized it up this way: "The major players have been the same for a long time, it is increasingly important to be a low-cost producer, and most participants have learned to sell and distribute their products efficiently." Some executives within the industry itself concur that it is maturing, and they express skepticism that semiconductors is a business in which additional start-up companies can compete and survive. An executive with a major semiconductor producer predicted that the venture capitalists will "get burned" on financing new semiconductor firms. In obvious disagreement with these views, other semiconductor executives generally see the industry as one still in its growth stage with plenty of room to accommodate new entrepreneurial endeavors. Their conclusion is based on rapid new product development, two to three dozen start-up companies in the last four years, and an industry growth rate of several multiples of what is normally associated with a mature industry.[9]

The point is that the *rigid* bell-shaped product life cycle curve is of little value to managers in high-tech companies. One cannot be certain whether products subject to rapid technological obsolescence follow a normal-curve pattern. And, even if they do, it is extremely arguable and judgmental as to when one product life cycle stage is yielding to another. As the semiconductor industry exemplifies, experts cannot agree among themselves.

Nonetheless, the *overall theme* or concept of the product life cycle can be of considerable strategic value to high-tech executives. Products do indeed experience life cycle evolvement from introduction to obsolescence and managers need to be alert to the general signs, say, that a growth phase is beginning to give way to more mature conditions or vice versa. The product life cycle concept of market evolution is fundamental to the formulation of sound marketing strategy. It is only rigid adherence to the idea of the bell-shaped curve with distinct phases that causes implementation problems for managers. The semiconductor case clearly illustrates that life cycle stages do not compartmentalize and that elements of growth and maturity can be present simultaneously.

What high-tech marketers need to watch for are oscillations between supply-side and demand-side market conditions. In the former, an innovation's acceptance is growing rapidly. The innovation is likely to be so useful that it is, in effect, creating its own market—for example, inexpensive, monoclonal, antibody-based diagnostic tests for various diseases. And, because of so much growth, competitors are not too concerned with their relative market shares. Then, as growth starts to slow or growth remains high but more competitors emerge, infighting and market-share battles appear. Eventually, this demand-side maturity brings about a shakeout of weaker competitors who can no longer compete. (Most frequently, those shake outs are technically competent firms with no marketing expertise.) Suddenly, there may be a significant technological breakthrough or new application of existing technology that causes the industry to oscillate back to a high-growth supply-side phase.

The traditional product life cycle concept suggests market conditions and sales volume as means to ascertain when a product is growing, maturing, and obsoleting. But, as discussed previously, sales volume of high-tech products can be misleading when used as a precursor of life cycle evolution. The revenues or units sold of a product associated with a particular technology typically reach an apex *after* a competing superior technology is introduced that ultimately renders the old product obsolete. Consequently, an executive attuned primarily to sales volume performance could easily draw the false conclusion that a product is still healthy when, in reality, its economic and technical *coup de grace* is imminent.

So, in addition to general market conditions and sales volume performance, astute high-tech executives will also closely monitor technology life cycles. A product's rating on the most critical measure(s) of technological performance, vis-à-vis competing products utilizing incipient and potentially superior technology, is far more important and certainly provides a much better technical/economic early warning system. In point of fact, sales volume performance is often no early warning system at all. It is usually a lagging indicator of economic and technical demise of a product.

High-tech executives who watch for supply-side/demand-side oscillations by tracking evolving market conditions, sales volume trends, *and* technology life cycles are prudent strategists. Leave one of the three out, however, and prudence is no longer present. A combined market conditions/product life cycle/technological life cycle approach will lead to more effective marketing and technical decisions. For example, as supply-side markets evolve and mature, entrepreneurial managers will need to be supplemented heavily with marketing executives experienced in "marketing warfare." In introductory market phases, there is usually enough growth to accommodate most of the pioneering competitors, and entrepreneurial-type executives thrive. There is nearly undivided focus on technical competence and inventiveness. But as markets mature, competitors intensely war with one another for share and, as a result, executives with different orientations and expertises are required. The emphasis in the market will have shifted away from technical innovation to distribution, advertising and sales promotion, and pricing. When confronted with ever-intensifying competition in personal computers, particularly from IBM, Apple Computer hired as its president a prototype consumer-products marketing executive from Pepsi. How did Digital Equipment Corporation (DEC) react, albeit belatedly, after it became evident that the company's internal preoccupation with technological preeminence in minicomputers had cost it dearly in the lucrative personal microcomputer market? It launched a massive organizational overhaul to transform the engineering-driven firm into a marketing-driven competitor. DEC's president explained that product and marketing shortcomings, not technical ones, prompted the sweeping overhaul.[10]

The New Product Process

Without the benefit of "real world" data from a cross-section of firms in various high-tech industries, there is the temptation to reason that the new product process differs considerably from industry to industry. But, is product development and commercialization in biotechnology or pharmaceuticals unique and does it differ markedly from the processes used in robotics or electronics? There are differences here and there. However, contrary to intuition, the new product *process* is essentially the same for all high-tech industries and successful firms—once the process is boiled down and all the jargon peculiar to each high-tech field is eliminated. When we compared the new product process from company to company and across industries, we found remarkable similarities, not dissimilarities. Our data and findings pertain to high-tech companies with proven records in the marketplace to show for their efforts.

Which Corporate Functions Do Well-Managed High-Tech Companies Involve in New Product Development?

Virtually without exception, the chief marketing executive is the person most in charge of product development. This individual is seen as the corporate officer most responsible for new product development—from idea conception through to commercialization and beyond. This fact is revealing: Those successful market warriors are not engineering-driven, with marketing relegated to a subsidiary role. In the best high-tech companies, marketing's authority and responsibility extend well beyond the selling of products that the technical side of the business has developed. In market-driven high-tech companies, the first question asked is: "What is the extent of the present or latent future market need or desire for our potential new product?'' Which is quite different conceptually from asking: "Can we 'sell' the product that we may have the technical capability to develop?''

Other corporate functions and executives that market-driven high-tech companies heavily involve in new product development are these: the top R&D managers, the firm's chief executive officer, and, but to a much lesser degree, the top engineering and/or production manager. About half the time, the chief financial officer is also *integrally* involved. Legal personnel sometimes participate but, in most instances, only in a purely legal role. Their counsel on general management as it pertains to new products is rarely sought.

Genetic Systems Corporation

Product Strategy in a Leading Biotechnology Company

The company's product strategy focuses on short, intermediate, and long-term goals, designed to assure the introduction of a steady stream of new products throughout the next decade.

The first generation of new products is meant to achieve short-term goals: diagnostics for infectious disease and cancer. The company's microbiologists are developing products for prompt diagnoses of a diversity of human infections, including sexually transmitted diseases such as herpes, respiratory diseases such as pneumonia, and infections such as staph, that further weaken individuals that are already hospitalized for long periods of time.

The second generation of new products is intended to accomplish intermediate-term goals: automated blood cell typing. The company's immunobiologists are working on products for the automated typing of human blood cells. These products will be used to determine compatibility for blood transfusions and organ transplants, susceptibility of persons to various diseases, and paternity in court cases pertaining to child support and/or custody.

The third generation of new products is geared to attaining long-term goals: treatment of infectious diseases and cancer. The company's Therapeutic Department is developing monoclonal antibodies of human origin for the treatment of problematic bacterial infections that often occur in both hospitalized patients with long-term illnesses and burn victims. Scientists in the Oncogene joint venture (with Syntex Corporation) will strive to produce human antibodies that are therapeutic for the most common cancer forms occurring in the United States—leukemia and cancers of the breast, lung, prostate, and colon. Studies will also be made of oncogenes, the genetic factors responsible for the cause of cancer, and cell growth factors that influence the behavior of cancer cells.

Information adapted from the Genetic Systems Corporation 1982 Annual Report.

Table 5-1
Where Successful High-Tech Firms Get New Product Ideas

The Most Productive Sources (in order of importance)	*Other Productive Sources*	*The Least Productive Sources*
Asking customers	Top management	Inventors
R&D	University laboratories	Commercial laboratories
Unsolicited customer suggestions		Industrial consultants
Watching competitors		Patent attorneys
Company sales representatives		

Where Do Leading High-Tech Companies Obtain Their New Product Ideas?

Listings of productive and not-so-productive sources for new product ideas are included in table 5-1. Among a broad spectrum of high-tech companies, few in the start-up stage, we found that the most productive sources are firms' present customers—the answers obtained by asking customers about their needs and preferences. So, still another time, it becomes evident that product development is market-driven in these successful companies. *Asking customers* is used broadly; it includes person-to-person approaches to industrial customers as well as the more impersonal approaches used by formalized marketing research in consumer-goods markets. Asking can be undisguised or it can be veiled to get at latent needs that the prospective customer is largely unaware of or otherwise unable or unwilling to articulate.

The list of most productive sources encompasses research and development, unsolicited customer suggestions, watching competition (especially by attending trade shows), and company sales representatives. Four of the top five most productive sources of new product ideas for high technology are market-driven—asking customers, unsolicited customer suggestions, watching competition, and company sales representatives. Moreover, their usefulness confirms that a company's own sales force is its best arm for market intelligence gathering. If not a company's sales reps, who is in a better position to ask customers and to receive their unsolicited suggestions and to watch competitors?

Additional productive sources for new product ideas, but not as useful as the top five, are members of higher corporate management and university laboratories. Although they receive some support from high-tech firms, inventors, commercial laboratories, and industrial consultants are generally not thought of as good sources of new high-tech product ideas. And patent attorneys receive no support at all in this regard.

We believe it critical to emphasize that start-up companies in extreme supply-side markets (for example, gene-splicing) are likely to have a different set of new product or process idea sources. These notably include the entrepreneur/scientist's own patentable breakthrough or the new application of an existing discovery to a new field.

What is the General Process that New Products Follow in Leading High-Tech Companies? And What is the Role of Marketing and R&D in This Vital Process?

As with any new product, the process initiates with an idea and concept testing. Thereafter, in high technology the sequence follows that given in figure 5–4, from identifying prospective customers to test marketing (in about half the cases), and, finally, on to commercialization.

Winning new product practices and philosophies emerge from looking at the way leading high-tech companies view and go about their businesses. First, they do not first develop new products and then worry later about how to market them, although they do feel that a "develop now and worry about marketing later" mentality accurately characterizes many high-tech companies, perhaps most. The proven companies also say that product planning is not essentially an R&D function. This precept is backed in practice by the heavy involvement of non-R&D personnel in the new product development process from beginning to end. Notice in figure 5–4 that both marketing and R&D are involved in the formative stages. Second, demand estimates are considered at best to be ball-park estimates of market potential and are not sophisticated projections normally associated with mature markets where there is much more of a past history to go on. Third, leading high-tech firms are calculated risk-takers. There is a noticeable lack of test marketing: A product is marketed in a limited geographic area or with a small number of buyers before a full-fledged commercialization effort is made. In many market situations, particularly in mature consumer products industries, test marketing is the rule—almost an ironclad rule. In high-tech industries, test marketing is neither the rule nor the exception. Approximately half the proven high-tech companies in the United States test market before full-scale commercialization of a product and half do not. As indicated in chapter 4, test marketing of high-tech products runs the risk of providing competition with the time and technological information it needs to react. Technological secrecy and surprise are key elements of innovativeness.

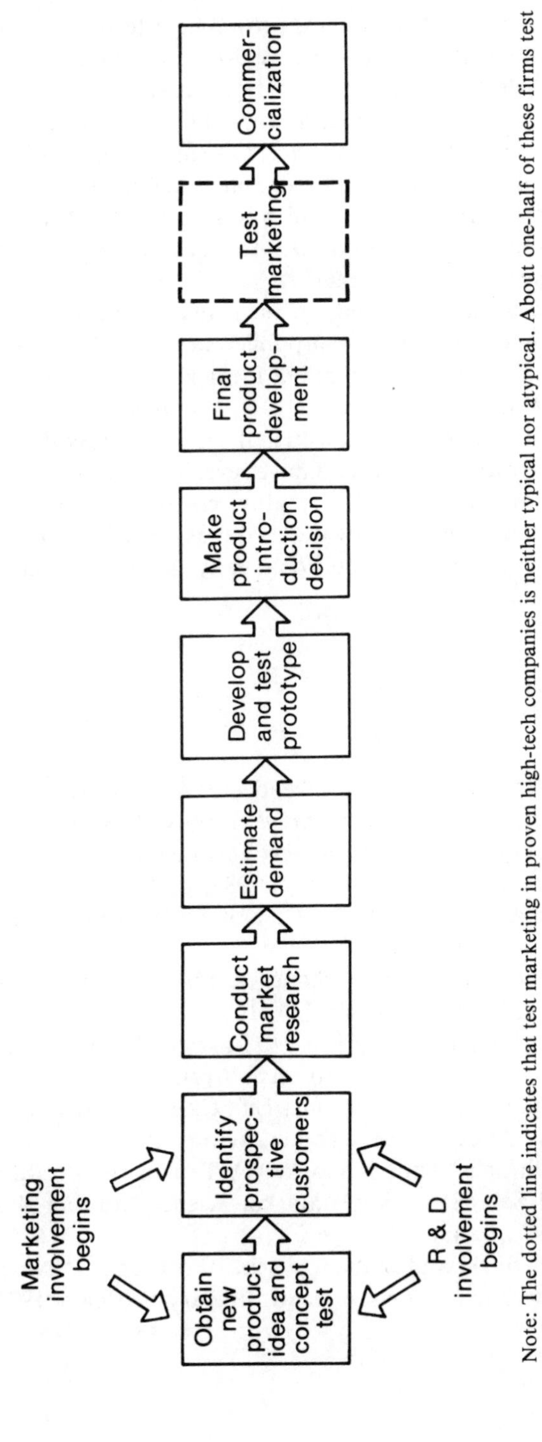

Note: The dotted line indicates that test marketing in proven high-tech companies is neither typical nor atypical. About one-half of these firms test market and one-half do not.

Figure 5–4. The Typical New Product Process in the Successful High-Tech Firm

Even though they are risk-takers, these proven high-tech companies are not run by the foolhardy. Reference to figure 5-4 will show that the companies rely on a great deal of customer identification, demand forecasting, and marketing research. A vignette demonstrating the exact opposite—needless risk-taking—comes to mind. It concerns a small high-tech firm that has been in existence for several years, barely surviving on venture capital. It has yet to show a profit for any year, which is little wonder. Without any professional marketing research (according to management, it cost too much), the company decided to market an electronic device which supposedly measured voice stress and, in turn, overall body stress. Not long thereafter the company was in well-publicized trouble with the U.S. Food and Drug Administration for making medical claims without FDA approval. To compound the problem, the U.S. Postal Service questioned the company about making the claims through the mail. If the company had doubts, a pretest of its advertising would have revealed that the advertising copy had medical connotations, and a lengthy, costly, and embarrassing incident could have been averted. All this brouhaha was over a product whose market success was deemed questionable by the most optimistic of objective estimates.

Notes

1. Arnold C. Cooper and Dan Schendel, "Strategic Responses to Technological Threats," *Business Horizons,* February 1976, pp. 61–69.

2. Richard N. Foster, "To Exploit New Technology, Know When to Junk the Old," *Wall Street Journal,* May 2, 1983, p. 22.

3. Leslie G. Cook, "To Innovate—Or Not To Innovate," *Research Management,* May-June 1983, p. 7.

4. Walter Kiechell III, "The Decline of the Experience Curve," *Fortune,* October 5, 1981, pp. 139–140, 144, 146.

5. Susan Chace, "Japan Trying Hard to Take the Lead from U.S. in 'Intellegent' Computers," *Wall Street Journal,* April 4, 1983, p. 15.

6. John E. Smallwood, "The Product Life Cycle: A Key to Strategic Marketing Planning," *MSU Business Topics,* Winter 1973, pp. 29–35.

7. Rolando Polli and Victor J. Cook, "A Test of the Product Life Cycle as a Model of Sales Behavior," Market Science Institute Working Paper, November 1967, p. 43.

8. Nariman K. Dhalla and Sonia Yuspeh, "Forget the Product Life Cycle Concept," *Harvard Business Review,* January-February 1976, pp. 102–112.

9. "Chip Wars: The Japanese Threat," *Business Week,* May 23, 1983, pp. 80–90.

10. "A New Strategy for No. 2 in Computers," *Business Week,* May 2, 1983, pp. 66–75.

6 Directing the Presentation

How important is a high-technology company's presentation (i.e., its communications links with customers, potential customers, and others) to its marketing success? Nowhere is the need for marketing expertise more evident than in the high-tech firm's total communications efforts. The high-technology company's success (and yes, even survival) depends not only on its continued ability to develop new products and processes (its R&D component), but also on its ability to make its target markets aware of their availability and quality.

Diffusion-of-innovation theorists (see chapter 4) too often seem to down-play or ignore entirely the importance of marketing in general and promotion in particular in determining the rate at which a new product or process gains market acceptance. Yet, practitioners typically begin to track an industry's real growth and development from the time that the companies with proven marketing skills enter the competitive picture, which was illustrated, for example, when General Electric Company entered the robotics field and IBM and Texas Instruments were seen standing in the wings. *Industrial Marketing* (now *Business Marketing*) reported this entrance by calling it the point at which the industry stepped into the big leagues. The reason given was that "those companies [GE, IBM, and TI] hold something even more precious [than technological know-how] in the fledgling industry—marketing muscle."[1] Their vast marketing expertise was seen as overwhelming anything presently in the robotics field.

To carry this line of thought a bit further, look at what others have said about the need for and use of market muscle, which we can partially define as communication strength. First, Roger Bennett and Robert Cooper, in a *Business Horizons* article, have cited marketing muscle as being such a significant success factor that it has led to the current emphasis on a market-oriented R&D strategy. This strategy has produced a situation in which this country "is spending a relatively constant percentage of its GNP on advertising, while the proportion devoted to R&D is falling."[2] Since we have stressed the supply-side marketing orientation in this book, we obviously do not recommend a nonproduct (R&D) strategy (nor do Bennett and Cooper); but we do emphasize the importance of marketing and, particularly, advertising.

Second, to go one additional step, a well-known financial analyst and robotics specialist was quoted as saying "marketing is at least as important in this stage [of the robotics industry] as technology or financial strength."[3] And the senior executives of Schering-Plough, Fairchild Industries, United Technologies, Gould, and similar high-tech companies continually cite marketing capabilities, as well as R&D strengths, when describing their company's future potential.[4] We could go on, but a single theme recurs—the need for a balance between successful product and/or process development and effective marketing and communications efforts.

Marketing-Advertising Confusion

Edward G. Michaels, a McKinsey and Company, Inc., director, recently noted that many business people equate marketing with advertising. He says that if you randomly select one hundred managers "from non-packaged-goods consumer product companies and ask them for a definition of marketing," more than half would say, in effect, advertising.[5] We agree, but not just because of the hype surrounding advertising.

Not only is advertising and, more broadly, communication the most visible arm of the marketing mix, it is also the arm that is most frequently asked to be the "bell cow" of the firm's marketing strategies. Increasingly, communication, whether it be through public relations, advertising, or sales promotion efforts, is expected to:

- provide customers or potential customers with information
- prescreen the potential customers via the media selection
- provide leads by stimulating interest in the product or process
- assist in forming and/or enhancing the company's image or reputation
- increase the public's and prospective customers' awareness of the company and/or its products or processes
- assist in motivating employees by fostering intracorporate confidence
- be a major cornerstone in the firm's positioning efforts
- increase the value of a corporation's stock
- improve the salespeople's chances of talking directly with the prospect
- complete, in some instances, the transaction or sale

Granted, there is some redundancy here, and we will see shortly that not every advertisement or public relations message is directed toward all these objectives. Yet, advertising's role potentially cuts across all the other marketing efforts of the firm—in a supportive fashion. We, however, will employ an even broader term than communications to describe a key marketing function—the informational and sales message linkage between the firm and its target publics.

Why Presentation?

Do we see presentation as merely a synonym for communications or do we perceive it to be more descriptive of the multiple information/selling related activities of the high-tech firm? And if we do see presentation as going beyond communications, in what ways does it differ? The answer is that communications has come to be defined or used too narrowly, otherwise we would have found it to be more acceptable for our discussion of high-technology marketing. Communications is increasingly being used merely to describe the firm's advertising and public relations activities.

As we define the term, presentation embodies all the planned external and internal contact (information dissemination and nonpersonal sales efforts) that occurs between the company (represented by its management) and its targeted publics. The sorts of targeted publics that the firm may wish to contact plausibly include its customers and prospective customers, its shareholders, the broader financial community, key governmental officials, its distributors, and even its own employees. This contact can range from the use of advertising in a trade journal to personal contacts by the company's sales force at a trade show. In every instance, the contact needs to be carefully planned so that it is consistent with the desired objective of quality the firm seeks to project.

For instance, an important point of interaction between the high-technology company and outsiders is often the trade show or fair. In preparing for a spring 1983 trade show in New Orleans, Novo Industri used advertising in trade journals; sent direct mail to key customers and influencers to invite them to visit the company's display; built a display that was tightly coordinated (and used the same theme and creativity) with its advertising; and involved its sales force and distributors at the display itself. Not surprising, you say? Well, the point to emphasize is that the total trade show effort of this Danish company was tightly planned and orchestrated so that its entire informational effort—its nondirect selling contact with its publics—would maintain the quality objective it desired. No public relations announcement, available brochure, or whatever varied from the way the company wanted its offering or message presented.

As an experienced marketing communicator responding to one of our high-tech surveys so aptly put it, "the issue [promotion] is one of staging the presentation of information from nonpersonal through personal sales tools to elicit an understanding of our purpose or impact in the marketplace (more than product features) and increase the propensity of an account target to engage our resources in a long-range working relationship.'' This biotechnology company executive is well aware of the importance of the presentation; he correctly sees the magnitude of the total communications effort.

Reputation

Since the concept of the presentation is so essential to our discussion throughout the remainder of this chapter, let us look again at points we have been making, but this time from a slightly different perspective. When one speaks of high-technology firms or industries, he or she thinks in terms of quality, leadership, R&D excellence, and a host of similar company and product attributes. Some firms claiming these attributes have high credibility, while others do not. It is interesting that an attribute for success that is mentioned frequently is reputation. Philip Maher recently described servo controlled robots as the ''Rolls Royces'' of the industry,[6] and the readers knew what he meant by Rolls Royces.

It was noted earlier that when GE entered the robotics field, the industry joined the big leagues. Why? Was it because GE has enormous resources and strong R&D—partly, of course. But, it was also because of GE's reputation—a reputation that has been nurtured by decades of communications efforts to inform its publics of its quality.

Consider the following illustrations dealing with the perception of quality and excellence:

> TRW's massive advertising campaigns in the United States during the past few years that have successfully enhanced its reputation as a high-technology company;
>
> IBM's major corporate campaigns that spanned Europe during the late 1970s and early 1980s focusing on its R&D capabilities;
>
> the participation of several Japanese high-technology firms in a special 27-page advertising section of the July 18, 1983, issue of *Business Week,* devoted to enhancing Japan's reputation for high-technology leadership.

These companies are aware that reputation comes not only from production and R&D excellence, but also through presentations informing their publics of their excellence. And given the importance of reputation to the high-tech company, it would be difficult to overstate the need for management to plan carefully all the high-technology company's nonpersonal and personal communications.

Personal Selling versus Other Contacts

As figure 6–1 depicts, the high-technology company, like other marketers, tends to follow two major communications routes in contacting its target publics. Through its sales force or representatives and distributors it makes personal contacts, while through advertising, public relations, and sales promotional materials, it achieves nonpersonal contacts. However, especially in business-to-business marketing, for instance, during a trade show, the two methods of communication may converge.

While we have chosen to separate our discussion of the general area of nonpersonal communication from that of personal selling and sales contacts, it is essential to emphasize that the two are inextricably linked together. Everything from the demeanor of the salesperson or representative to his or her product knowledge and statements about the high-tech firm's products or processes must be consistent with the company's nonpersonal contacts with the same publics. And this is true regardless of whether we are talking about a biotech firm in a narrowly defined business-to-business marketplace or a Texas Instruments that must reach a broad spectrum of industrial and consumer markets.

It is the company's chief marketing executive's responsibility to ensure that the two avenues of communication to the target publics are in concert; or, in other words, that its nonpersonal presentation and its personal selling efforts are consistent. Otherwise, the results can be disastrous, as the following example demonstrates.

A highly reputable manufacturer of consumer electrical appliances was stressing superior quality in its national advertising, but in a key southern city its primary outlet was a somewhat questionable discount store. This inconsistency was not lost on its potential market, although for some reason it had not been recognized by the company's sales management team. It was only after the company's market share in the city dropped sharply that the firm—aided by a consultant—caught its error. A company's target publics will be confused and turned off by inconsistencies. Therefore, the high-technology firm needs to establish formal control mechanisms to ensure that its personal and nonpersonal communications are in line. This control must begin with the establishment of the company's marketing objectives, be followed up with consistent personal selling and communications goals, and continued through every contact that is made with its target publics.

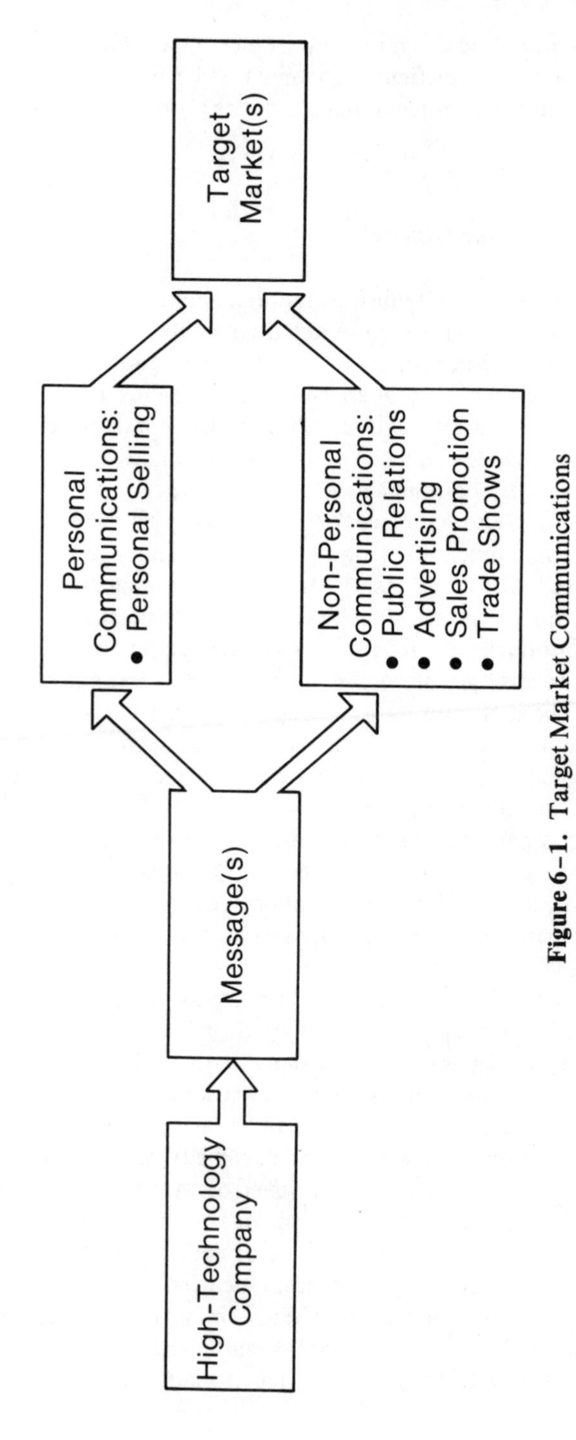

Figure 6–1. Target Market Communications

About This Chapter

Having stressed the importance of the presentation concept and the need for planning in all phases of the presentation, the remainder of the chapter is devoted to specifics—to implementation strategies and tactics. Naturally, there is a wide variation in the presentation requirements among firms and industries. With our broad-brush approach, the marketing of high-tech products and processes must somehow speak to the problems of a robotics or software firm just reaching $1 million in total sales as well as the major multimillion- or billion-dollar companies in the pharmaceutical or computer hardware fields.

But, as we have said, all high-tech firms, regardless of size, age, R&D expenditures, or market share, do have certain concerns in common. There are activities that need to take place regardless of the company's scale of operations. These activities provide us with many of the topics included in the remainder of the chapter. For convenience, the topics are stated in the form of questions that need to be answered.

What do we mean by planning the presentation, and what form might it take?

What are realistic presentation objectives for a high-technology firm?

What kinds of communications media are available and used by high-technology marketers?

Is corporate advertising an essential part of the high-technology company's presentation?

What about activities such as budgeting, control, and advertising research? Can the effectiveness of the overall presentation be determined?

What should be the role of the advertising agency in the high-technology company's presentation efforts?

Why are sales promotion and, especially, public relations key components of the high-technology firm's presentation?

There are corollary topics that also merit comment, mainly the recent development of the high-tech advertising agency and the rapid evolution in Europe of the single, integrated consumer market. We begin with the most basic or fundamental step, the development of a plan for the company's presentation.

Planning

Naturally, planning is a continuing activity in firms of all sizes and involves each function and several levels within them. Just as the marketing plan for a predetermined time period must be related to the corporation's overall plan, the presentation plan must be tied to the achievement of the objectives assigned to marketing.

Robert Kriegel has succinctly described the activities of the sales and communications departments in this respect. He says that from the marketing-related elements of the company's corporate strategy (corporate strategic plan),

> marketing sets its tactical plan, which in effect establishes the objectives of its sales and communications programs. The communications department assists, by writing measurable time-dimensioned objectives according to the tactics specified in the marketing plan. Communications executives then choose strategies and tactics to achieve their objectives. [Sales management similarly will set objectives, strategies, and tactics—some closely related to the communications plan—in line with the (company's) marketing plan.][7]

Although Kriegel is writing about the industrial marketer in general, the points he makes are essential to the presentation planning process for a high-technology firm. He correctly notes the tight linkage, for example, that is needed between the personal selling and communications plans. One is reminded that all "of the firm's advertising activities should be consistent with the objectives that have been established for advertising and should be consistent with and complement the company's total marketing efforts."[8] Failure to do so has caused many firms "not only considerable embarrassment, but also unnecessary costs."[9]

In the same article, Kriegel also indicates five key elements in the communications plan: "(1) objectives; (2) message and positioning strategies; (3) specific action plans (tactics); (4) measurement procedures; and (5) budget."[10] We discuss each of these elements at some length later in this chapter, but Kriegel's inclusion of measurement procedures and the budget as part of the planning process bears special attention now. Too often the planning process has been seen as merely establishing objectives and then developing strategies and tactics to achieve them. This concept is quite unrealistic and, from the standpoint of the presentation planner, even unprofessional. Every attempt should be made to evaluate the performance of the various elements of the presentation and to consider this performance in light of budgetary limits. This activity must begin at the planning stage, as suggested here, and requires the establishment of measureable objectives and a corresponding budget—and their subsequent measurement.

Setting Objectives

We indicated earlier a rather heady list of objectives that presentations have been called on to achieve. While a single firm might attain all of them over the longer term, it would be delusory to expect an annual advertising campaign or, possibly, the campaign plus a complementary public relations and trade show or sales promotion effort to achieve any one of them fully.

Most presentation objectives, in order to be measurable, relate to awareness, image, sales, or market share. And this is so whether we are focusing on domestic or international presentation objectives. For example, a recent book dealing with the management of international advertising, notes five realistic measurable objectives, and each involves either increasing the awareness of the product, improving the image of the product, increasing the product's sales, or building its market share.[11] The book's authors caution against using changes in sales or market share as the only measure, unless the possible effect of other elements in the total marketing effort is also considered. They point out that sales and market share are frequently influenced by other marketing factors, for instance, the product's quality or its price.

What would be considered realistic presentation objectives, recognizing, of course, that they must be in line with the firm's marketing goals? It has been said that one of the problems of the high-tech marketer is that it must deal with a bifurcated market—it must effectively reach both the specialist (scientist or engineer) and the generalist (nontechnically trained company official). An executive of Cincinnati Milacron, Inc., made just this point when discussing the company's exhibit at the National Plastics Exposition in 1982.[12] Given the problem, a company producing a rather complex software application might establish the following presentation goals: to increase the awareness of senior executives in its target market as to how an application might benefit their company and to indicate to the same companies' technocrats the special process of the software application.

Now, in order to make this objective more measurable, it would be preferable to conduct basing point research to ascertain how many CEOs and presidents in companies in the target market were aware of the need for or benefits of the software application prior to the firm's presentation efforts. A follow-up study with representatives of the same group would then allow the company to identify the magnitude of change resulting from the presentation. Say that the prepresentation findings suggest that 10 percent of the CEOs and/or presidents were aware of the software application's benefits. A realistic goal might be to increase this figure to 30 or 40 percent during a one-year period. Without doubt, the "clearer the objective and concomitantly the better the audience is targeted, the stronger the advertisement or campaign and the more likely its success can be measured."[13]

Similar to the software example, a company in pharmaceuticals might find that its target market is not aware of its entry into genetic engineering, perhaps through an acquisition or a licensing agreement. Or, possibly of even greater concern, the company's image among CEOs and presidents relative to other firms in its industry may be such that it is not perceived to have significant research clout. Thus, any announcement of subsequent breakthroughs or new applications by the company would be ignored or have limited credibility. Again, an approach like the one followed by the software firm involving prepresentation and postpresentation research would be recommended in order to measure whether the company's presentation efforts were successful.

To employ a sales or market share objective, the company needs to follow a somewhat different tack. For, in essence, its presentation objective in the short term at least would be virtually the same as its overall marketing objective (or, under a product line or divisional structure, it would be synonymous with the marketing goals for that product or division). Thus, the marketing objectives and the facilitating presentation objectives should be based on information pertaining to the projected growth for the industry, the expected actions of competitors, various economic forecasts, and the like. To measure the effectiveness of the presentation, it would be necessary to minimize other marketing activity changes. For example, if the firm were to make significant price or product changes, it would be difficult to assess the effect these changes had on the presentation's success or failure in meeting its objectives. Typically, when sales or market share presentation goals are established, the objective is stated more in direction than specific dollar or percentage amounts. The presentation objective might be stated, for instance, as "a move from fifth to fourth place in market share in the industry." This approach tends to allow, at least partially, for the impact of the other marketing variables and the overall industry environment on the final sales or market share picture for the year.

Study Respondents' Views on Objectives

The respondents in a study we conducted among the best-known advertising executives in a broad range of high-technology industries provided their views on advertising objectives. They were presented with a list of possible goals and were asked the frequency with which each had been an objective of their firm's advertising. The list included:

to obtain inquiries;

to confirm markets;

to identify markets;

to increase corporate image;

to announce new products;

to increase awareness of the company;

to communicate product attributes; and

to increase trade show display attendance.

The respondents converged on two major types or classes of objectives: (1) announcing new products/obtaining inquiries and (2) communicating product attributes. In addition, most of the respondents believed awareness to be at least a secondary objective, and a few considered increasing image to be one of their "often used" goals. Other objectives received little or no attention.

Consumer advertising tends to have the objectives of communicating attributes and increasing awareness or image, while industrial or business-to-business advertising has traditionally sought to stress new product announcements and obtain sales leads. A current view suggests that the typical objective of the high-technology firm's advertising lies somewhere in between—offering a challenge to those attempting to develop a creative strategy to implement the objective. This challenge is greatest for firms in high-technology fields where dual markets exist, or where the level of user sophistication is so great that it is much simpler to stress emotion or to deal with awareness and image. Some even hold the view that the new breed of scientists and engineers—the ones that compose many target markets—react more like consumers (emotionally) than like the "old line" industrial buyers (pragmatists). As Richard Reiser, a high-tech advertising agency head, told us, "The high-technology customer is unfettered by tradition; that is, what are perceived to be objective reasons. The high-tech buyer is only fettered by what is 'in'." This thinking led, for example, to a successful campaign in which his agency gave a computer (RT Screen) a personality—"the dumb terminal"—à la Volkswagen "bug."[14]

In any event, it is clear from the study that several different advertising objectives are being widely used and that the need to increase awareness or to enhance image is a well-recognized or oft-employed advertising objective, or at least a secondary objective.

Creative Strategy

Once the presentation objectives are finalized, the marketer enters the implementation phases of the planning effort. The first phase requires the selection of an overall theme strategy that is *explicitly* designed to achieve the presentation goals. This sounds simple, and it is so rational that few would argue with this approach. Yet, how many advertisements have we seen, for example, that clearly seem to ignore this point? For examples of poor or problem industrial advertisements, it is helpful to follow the "Copy Chasers" section found each month in *Business Marketing*. An evening of television viewing will offer rich illustrations of poor advertising in consumer marketing.

Next, we issue an important caveat: The high-technology marketer must recognize that there are few, if any, mystical qualities associated with creative theme development. The best creative efforts are those based on an intimate knowledge of the target market and its unique qualities, or even peculiarities, combined with sound research.

A first step is the development of a broad conceptual theme which will ultimately provide the basis for one or more creative alternatives, an overall theme strategy that will be carried forward in the company's advertising, public relations, sales promotion, and trade show presentations. What is entailed?

In chapter 3, the importance of carefully identifying the market(s) for the product or process was stressed. By carefully screening and establishing priorities, the corporation has positioned itself to focus on the industry or industries offering the best opportunities for new products. For an existing product or process, the company may even choose to pursue a specific niche within the market. If the positioning process was thoroughly completed, the presentation theme planners—most likely the advertising manager and his or her advertising agency account executive or supervisor—have a wide range of data available. On the one hand, they may have only secondary material on the industry, for example, size of firms, competitive positions, etc., while on the other hand, they could have data resulting from studies conducted by the company and designed to obtain information relating to prospective buyers' needs and characteristics. Irrespective of the presentation objectives, the planners should review this information in order to understand more thoroughly the target market.

The specific kind of information that is ultimately needed for the development of the theme strategy will vary according to the objective. If the objective is increasing sales or building a market share, then the theme planners probably will want information on buyer or prospective buyer needs and product or process usage. And there are other related questions. Will it be important to stress benefits alone or will it be necessary to provide comparative data? What kind of buyer concerns are most compelling? Is image or service in this instance more important to market share or sales than product attributes? If answers to questions like these are not already available, then those involved in developing the theme strategy would need original research focusing on the questions. Yet, in the case of some business-to-business industries, such as the construction field,[15] studies have shown that limited preadvertising or postadvertising research is done. Although we recognize that many firms have limited presentation budgets and would prefer to avoid gathering primary data, it is important to note that mistakes in presenting a company's message can also be quite costly in terms of wasted media expenditures and lost potential sales.

We again turn to the use of the focus group described in chapter 4 as a useful and relatively inexpensive research tool in theme strategy formulation. Through the interaction of a small group of representatives of the corporation's target market, considerable insight into appropriate themes can be uncovered. For example, it may be learned that a primary factor inhibiting the sale of products or processes similar to what the company will be marketing is safety. If greater safety is a benefit the company's product can offer, then its overall theme strategy may have been determined.

Image and Awareness Themes

Later in this chapter, we consider the value of image advertising; however, let us look at image and awareness advertising objectives simultaneously for the moment, as both require somewhat different information for theme strategy than do the sales or market share objectives. Image and awareness require an indication of the target market's perceptions of the company either viewed alone, compared to other corporations and competitors, or both. In addition, it is useful to determine the reasons why a company has a poor image or awareness ranking compared to the competition.

Many marketing research organizations, such as Palshaw and Yankelovich, and Skelly and White, are active in the business of finding answers to these questions. Also, the company could make use of focus groups to indicate directions. A usual problem for the new high-tech firm is the lack of awareness on the part of its target market, while the "old line" firm may have a difficult time shedding its historic image in favor of the new high-tech directions it is currently taking. Based on published comments, it appears that a prime example of the latter is provided by Ball Corporation, a company with a good reputation in consumer glass products, but hardly a household word in space technology.

Time Period

Generally, the marketing objectives and therefore the advertising objectives are stated in increments of one year or less for both conventional consumer and industrial advertising. By contrast, high-technology markets are often so dynamic that even a year's time horizon may be unrealistic. The product diffusion process related to high-tech products can move so quickly that, in the first fraction of the year, the product or process needs to be introduced and, perhaps, explained. Later, very specific comparative attributes need to be stressed. For this reason, two presentation theme watchwords for the high-technology marketplace are flexibility and constant review. The former refers to the need to be able to adapt to change and the latter suggests a formal process to ensure that presentation themes are keeping abreast of the market's evolution.

Campaign Theme/Message

Once the presentation theme strategy has been developed, the specific campaigns based on the theme begin to take form. Recently in Europe, Digital Equipment Corporation took steps via corporate advertising to ensure that it was perceived as a leader, especially in technological innovation. This strategy translated into a central campaign theme, "We Change the Way the World Thinks," which appeared in all its advertisements.[16] Henkel KGaA, a major West German corporation, had a problem similar to that of the Ball Corporation. It was recognized as an important soap company, but desired a broader image more representative of its position in the applied chemical field. Its creative theme strategy was to use its employees' families to describe the diversified activities of the company. The exact wording varied, but in each instance the advertisement pictured the son or daughter of an employee describing what work his or her parent did at Henkel and then indicated the breadth and research depth of the company.[17]

In this phase of the implementation process, the advertising agency plays a prominent role. Given the direction provided by the overall creative theme strategy, the agency can provide several alternative messages that are designed specifically to execute the strategy. While the company advertising manager normally has the authority to select the one or more alternatives to employ in the actual advertising, it is generally helpful to go back to the target market by way of focus group or some other technique and determine which message or messages best achieves the desired theme strategy.

Presentation Tactics

Although we have frequently referred to the advertising manager in discussing the presentation planning process, there are others who should be involved. We would include those responsible for the public relations and sales promotion efforts, as well as the advertising agency account executive or supervisor, the trade show manager, and the marketing vice-president. (Although titles vary, we are making reference to those responsible for various aspects of the presentation.) In most firms, the overall authority for coordinating the presentation belongs to the advertising manager, since advertising accounts for the bulk of most firm's presentation budgets.

From the very beginning of the presentation planning process, all those involved should be constantly aware of tactics, juxtaposed against the salient objectives and strategy. This point is especially true in high-technology companies that are in the business-to-business or industrial field, because they know the limitations of the tactics they can employ. For example, a company may realistically be confined to direct mail and trade shows. One could generally make the same argument for high-technology consumer companies, since such considerations as the competition or the need to verbalize information may restrict their use of media. Some products, for instance, can best be explained and promoted in an audiovisual medium.

By tactics, we are referring to the specific implementation plan for presenting the select theme or message to the target market. Two tactical examples used by Kriegel in his hypothetical communications plan were:

> Mail brochure and letter to customer list in February 19XX and to sales department's "hit list" the same month. Distribute brochure in bulk to district sales offices and distributor list in February.
>
> Hire telemarketing consultant in January to set up incoming telephone program. Complete upgrade by March.[18]

The points he makes are important. The tactical aspects of the plan need to be specific in terms of actions and timing. Therefore, the tactical plans for implementing the presentation, whether it be a television campaign for TI, DEC, or TRW, or a trade journal campaign or a trade show for Mill-Rose Laboratories, should be specific, and the general times (May, June, etc.) stated.

But to reiterate, it should be clear why tactical issues are pertinent when the objectives, and especially when the strategies, are being considered. Furthermore, the same holds true for the budget, which is often small and may, in fact, be merely the residual of the operating costs. However, should this be the approach followed by the high-tech company? Should it dwell on its limitations when it does its presentation planning in general and its tactical efforts in particular? No.

Proactive versus Defensive Posture

A point that has been disappointing to us in several of our discussions with executives of high-technology companies is the extent to which they seem to see their presentation activities as being merely defensive in nature. Whether it is participation in a trade show or an advertisement in a trade journal, one too often hears the comment "I would not be there if it weren't for . . ." The reasons range from specific competitors to the view that "everyone would think we were 'out of business' or having financial problems if we did not appear." There may be some justification for the rationales they give; however, we recommend using the firm's presentation tactics in a proactive way. That means not just being there, but using a trade show or an advertisement to its maximum advantage. We can assure you from past experience that this attitude change makes a difference. For example, companies often use people hired locally rather than their sales representatives or regular company personnel for their trade show displays. The difference between just handing someone a brochure or recording their badge for future mailings and answering a critical question on the spot may produce quite different results. Running a routine advertisement in *Production Engineering* versus running a carefully targeted and quite creative advertisement stressing the theme strategy has similar implications.

Proactive Tactical Planning

We recommend a PERT-like approach in order to ensure proactive tactical planning, although any schema that permits considering all facets of the communications would be equally appropriate. What does this approach require? It necessitates the preparation of a grid chart for the year or the budget period. The various media and other elements of the presentation are arrayed at the side and the various weekdays of the year are shown across the top. Next, key dates can be entered if they are essential to the annual planning process. For example, the American Society for Cell Biology exhibition was held in Baltimore November 30 through December 4, 1982. If this were an important trade show, for example, one appropriate for the company's target customers or prospects, then it would be entered on the grid chart. Tied to this show would be direct mailings, perhaps new brochures and other sales promotion materials, advertising, and possible public relations news releases. The timing of these events would be indicated on the grid.

Referring back to the examples from Kriegel's article, it is important to develop specific tactical actions for the overall presentation plan. We believe that a system similar to the grid approach will assist in providing the needed controls; it is a way of assuring that the various actions are not only planned, but also are actually implemented at the appropriate time.

Besides Planning

What about the other nonplanning related questions that were raised earlier in the chapter? There are a number of specific topics that concern the presentation activities of a high-tech marketer. These relate to the kinds of media that are available and used by high-tech firms, the increased use of corporate advertising (and the whole question of image), the advertising research issue, the importance of sales promotion and public relations, and several issues concerning advertising agencies.

Media Available

What kinds of communications media are available and used by high-tech marketers? Virtually every possible advertising medium has been employed by some high-technology company, since this area of marketing involves both consumer and industrial or business-to-business products and services. In our surveys of high-technology industries, the most widely used media (in order of usage) were:

specific industry trade journals;

brochures (and other promotional literature);

direct mail;

trade shows;

general business magazines;

videotapes (sales promotion);

newspapers;

television; and

radio.

This order may come as no surprise. But the fact that there was substantial support for each medium might.

Trade Journals. The strong preference for specific industry trade journals is indicative of the fact that (1) the publishing field is very quick to respond to new technology and (2) trade journals provide a convenient way to reach a particular niche. Similarly, the high-tech consumer products producer has not been ignored by the specialized publication field, as seen by the plethora of home computer publications that have appeared on the magazine stands.

In carefully positioning an industrial or business-to-business product or process, a company tends to find itself with three extremely valuable media: specific industry trade journals, direct mail, and trade shows. However, the specific industry trade journal offers possibly the greatest ready access to the market. First, it does not require the firm to develop or acquire a mailing list—a somewhat formidable chore, particularly overseas. Second, if it is a publication of a professional society or is an audited business publication (BPA), its readership has been in effect validated or prescreened for the advertiser. And third, there is not the lengthy waiting period associated with trade shows for announcing a new product or process or for stressing new benefits. Small wonder then, that such high-technology trade journals as *Robotics Today, Biomedical Products, Laser Focus,* and the *Journal of Petroleum Technology* are successful. For example, in August 1982, the *Journal of Petroleum Technology* showed a marked increase in total advertising pages over the preceding year, which was not the trend for business publications in general. As indicated in figure 6–2, high-technology fields are well-represented among the ten fastest growing business magazines by advertising revenue, as reported by *Advertising Age.*

Television. Possibly the greatest future media opportunities for many high-tech companies lie with the more imaginative use of television. Whether it be cable television or videotaped sales messages to be shipped directly to customers or used at trade shows or video-telephone hookups or even international satellite television, this visual medium offers interesting possibilities for the high-technology firm.

Consider just a few possibilities relating to cable television alone:

> *Advertising Age* (June 13, 1983) reports the FNN, the Financial News Network that began in November 1982, has 625 affiliates and is reaching a total of 6½ million viewers.[19] (Does this network offer possibilities for high-technology firms that wish to build their image and/or enhance their awareness among the financial community?)
>
> The results of a ten-month study by J. Walter Thompson USA and Adams-Russell indicate the popularity of Cableshop, an "endless stream of three- to eight-minute informercials . . . as a browsing medium," used as people normally read catalogs.[20] What potential does this offer for both consumer and business-to-business high-technology advertisers, who often need longer message time to explain their new product or process?
>
> Videopath switched broadband network (which gives two-way interaction for microwave transmission of video and data, both live and

1. Computer Decisions	6. EDN
2. Mini-Micro Systems	7. Petroleum Engineering International
3. Institutional Investor	8. Computerworld
4. The Blood Horse	9. Oil and Gas Journal
5. Motor	10. Diversions

Adapted from Brian Rogers, "Gearing Up for an Active Future," *Advertising Age,* May 16, 1983, p. M-10.

Figure 6–2. Ten Fastest Growing Business-to-Business Publications (listed in order of annual percentage advertising revenue change)

taped), is used for program transmission, advertising, videoconferencing and other services.[21] Could the high-technology company use this versatile cable network for anything from sales conferences to video-formed focus groups?

Our questions posed above just scratch the surface as far as cable possibilities are concerned. Teleport, a telecommunications facility planned for New York by the Port Authority of New York and New Jersey, has been called a step beyond the potential of cable TV, and it, too, offers videoconferencing as well as other advanced communications possibilities.[22] Clearly, the high-technology marketer will want to keep abreast of television-related developments, which can have far-reaching effects on the company's ability to transmit its theme strategy to ever-better defined target customers and potential customers.

Other Media. Even though we have drawn the reader's attention to specific industry trade journals and television, other media also offer striking potential for the marketer of high-tech products. For instance, direct marketing, either via telephone or mail, is extremely important to both consumer and industrial or business-to-business high-technology marketers. For business-to-business advertisers, the telephone (telemarketing) appears to have moved ahead of direct mail in total annual expenditures.[23]

Trade shows, naturally, have a continuing place in the presentation tactics of most high-technology firms. Although they are not often classified among advertising media, they do have all the requisites—except the fact that the trade show is a form of personal rather than nonpersonal communications. Often the company's participation in a trade show can be included in an advertisement or advertising campaign in an effort to attract traffic (see figure 6–3).

Trade shows and individual trade show presentations run the gamut of sophistication and quality. The high-technology firm must establish priorities for the shows, as their entire communications budget could be spent on

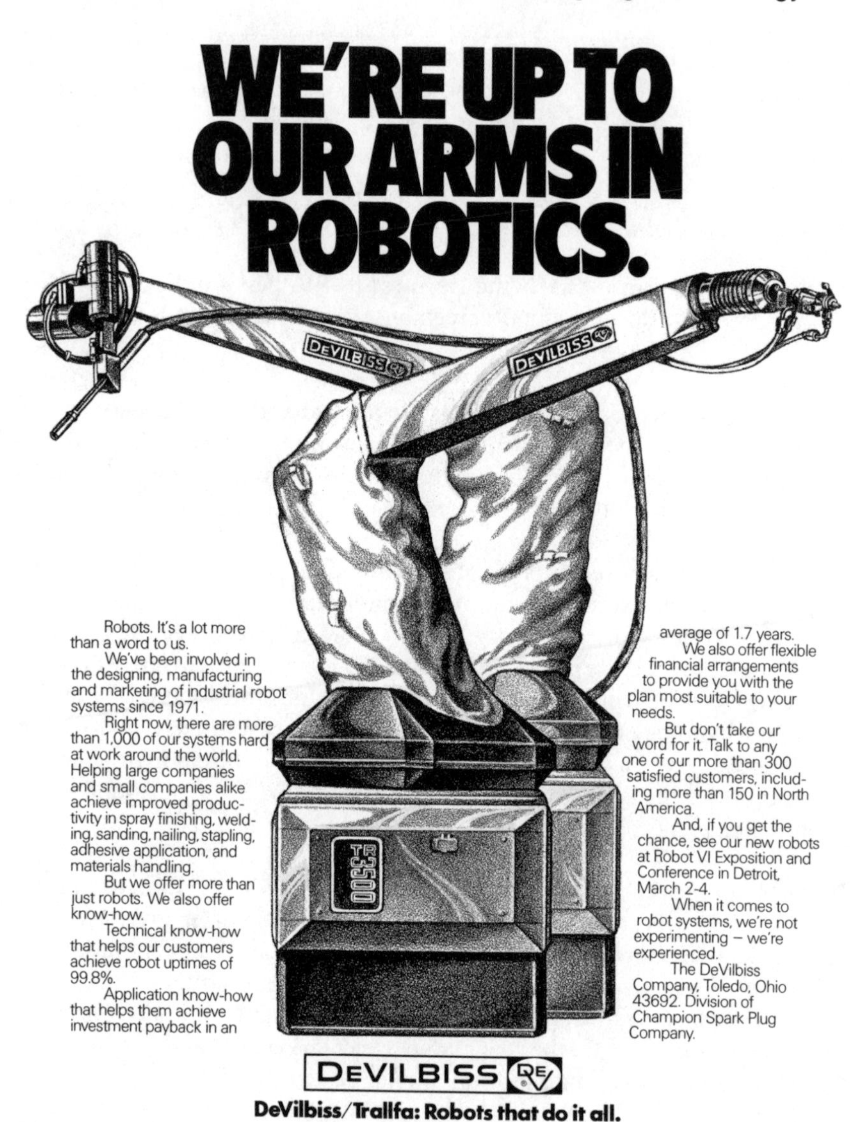

Note: The DeVilbiss Company advertisement appeared in the February 25, 1982, issue of the *Wall Street Journal* prior to the Robot VI Exposition and Conference in Detroit held March 2–4, 1982. The DeVilbiss advertisement was targeted toward the senior corporate executive and had two objectives. First, it was designed to stimulate interest in industrial robot systems on the part of both technical and nontechnical readers and second, to attract traffic from those attending the trade show. Reprinted with permission.

Figure 6–3. DeVilbiss Company Advertisement

this medium. It is important to account for audience duplication and to select the shows that come closest to matching the firm's target audience. Token participation might not be preferable to no participation at all, since any participation reflects on the company's image.

Corporate Advertising

Is corporate advertising an essential part of a high-technology company's presentation strategy? The term *corporate advertising* covers a range of different types of nonproduct advertising, including advocacy and controversy advertising. Our major concern is with corporate advertising designed to enhance the corporation's image, reputation, and level of awareness.

As we noted before, it is essential for a high-tech firm to have a quality reputation—one that is compatible with its level of R&D and fosters confidence on the part of its customers and potential customers. For example, does the firm have the credibility it needs if it is to have its new product or process accepted? It is well-known in the tire industry that the first company to develop the radial tire was not the first company successfully to introduce it. The originating company simply lacked the credibility to convince enough buyers that it had made the breakthrough.

Whether or not a company needs to have a corporate advertising campaign really depends upon factors like the company's size, industry, current image or stature in its field, volume of product advertising, relative level of awareness, competition, and current objective(s).

Two or three of these require some explanation. An industry itself may capture a high level of public interest, perhaps even create controversy, so that it develops a degree of notoriety for its leading firms. This ramification is seen today in the genetic engineering field, which has captured the public imagination. People can imagine all the possibilities the area of gene-splicing generates. Industry watchers recognize the leaders, and the business press relays this information to the broader business community. The *Sunday New York Times* ran a front page (business section) management rating of the three key companies in the industry, all of which were relatively unknown.[24] This publicity was especially good for Biogen, which received the highest rating. In the long run, which firms will maintain their image and level of awareness in this rapidly evolving high-technology field? A possible answer is suggested by the events in the robotics field when the GEs, Westinghouses, IBMs, and others entered—the attention shifted, to some extent, away from the pioneering firms.

Level of awareness and image are really the two sides of the sword hanging above the high-tech firm as it seeks to establish its reputation.

Some companies, a 3M, lack a consistent image, while others, an ITT, may have had to overcome negative images at one time or another. Similarly, a firm's level of awareness may be poor compared to its better-known competitors. Any effort to correct either problem, however, must be guided by the company's objectives and should be viewed in terms of longer-term payoff. While all advertising must be considered an operating expense for tax purposes, it is likely to be one of the major contributors to nebulous "good will" on the corporate balance sheet.

Interestingly enough, IBM adds a slightly different and perhaps more proactive view of corporate advertising: "IBM . . . considers its corporate advertising 'market prep' because it prepares markets for more specific ads." An IBM spokesman has been quoted as saying, "We stray from corporate to product advertising . . . it's a fuzzy border between sales promotion and reinforcing the corporate image."[25] Moreover, some companies are using "umbrella" campaigns that indicate their various product or process capabilities.

Viewed in the ways suggested and carefully targeted, a corporate campaign can play an important role in the company's presentation strategy. For this reason, TRW, DEC, United Technology, 3M, Sperry Corporation, and countless others have continued to employ corporate advertising programs. For a brief case history of a corporate campaign, read the vignette on page 000.

Advertising Agency's Role

What type of relationship should be fostered between the high-tech marketer and its advertising agency? A key decision on the part of the high-tech marketer is the determination of when to involve its advertising agency in the overall planning process. The advertising agency's role could range from merely implementing the company's strategies to assisting in the entire marketing planning process. At a minimum, most corporations use an advertising agency to assist with creative efforts, to handle its media selection and scheduling, and to perform similar tactical activities. This generalization holds even for large companies with their own in-house agencies.

The advertising agency usually has a much higher level of involvement. Our communications research indicated that the largest proportion of high-tech corporations do use an advertising agency in addition to having their own separate advertising department. Yet, relatively few included their advertising agency in their marketing planning activities. The agency was typically involved in the following activities (ranked in order of usage):

Managing an Imaginative Worldwide Corporate Campaign

Few, if any, high-tech corporations have been more effective in managing imaginative worldwide corporate campaigns than has Novo Industri A/S, Enzyme Division of Copenhagen, Denmark. Working through the Ted Bates agency network, Erik Borre, the International Advertising and Promotion Manager of Novo, has produced outstanding, uniform corporate print media campaigns. These campaigns are presented in the same standard format in each of the national markets where they appear. However, they are translated into the appropriate local languages—Chinese, German, French, Danish, Swedish, Italian, Spanish, Portuguese, Japanese, Arabic, Greek, and Serbo-Croatian, as well as English.

In addition, Mr. Borre has adeptly orchestrated these same campaign themes throughout the company's direct mail, trade show, and other promotional efforts. Mr. Borre indicates that his instructions to his local managers (and the local Ted Bates offices) include the following five worldwide campaign objectives:

1. to develop and fortify a concise joint corporate image
2. to obtain a uniform product profile
3. to obtain effective coverage through centralized media coordination
4. to get as many insertions as possible at the lowest possible costs
5. to secure identical creative standards worldwide

Several of Novo's most recent corporate campaign advertisements demonstrate just how imaginative and creative a corporate campaign for a high-tech company can be (see figure 6-4). Novo Industri is engaged in a wide range of biotechnological and other research. For example, its enzyme R&D departments have in-house expertise in gene technology, monoclonal antibody technology, etc.

Novo Industri A/S Annual Report 1982

1. creative (artwork)
2. creative (copy)
3. production
4. media scheduling
5. media selection
6. marketing research
7. marketing consulting

It is noteworthy that all the firms which employed advertising agencies used them at least for their creative work. Likewise a large share took advantage

Note: Reprinted with permission.

Figure 6-4. Novo Laboratories Advertisement

of their agencies' production and media related skills. Although a majority also utilized advertising agencies for marketing research, only a small proportion used them as marketing consultants. Part of the reason for the lower-usage-ranking for marketing research and consulting could be that many smaller agencies are not full-service agencies.

Some readers might question why agencies are so often used by high-tech companies which have their own advertising departments. Yet, this finding was expected because

the advertising departments are often small (two or three people);

the advertising agency offers (1) an outside perspective and (2) more specialization; and

the relative cost of using an agency is low.

The well-established advertising agency, in particular, has economies of scale on its side. Therefore, even though the strict agency commission system is becoming "outmoded, a mere vestige of past and simpler days. . . ,"[26] there are still cost, as well as expertise, advantages to employing the services of a good, high-technology-experienced advertising agency. The study findings corroborated that the bulk of the companies we surveyed were aware of this.

Besides the points already mentioned, one should not ignore that synergy can result from having several bright, creative people involved in the formulation of a campaign theme or that it is becoming especially useful to have an agency with foreign affiliates to permit the corporation to employ centralized planning for its worldwide promotional activities.[27] For the high-technology marketer, a measure of control over its foreign marketing and communications activities is essential, if for no other reason than to ensure the maintenance of promotional quality that is congruent with its desired reputation.

There has been a recent trend toward the development of advertising agencies specializing in high-technology firms or industries. Much of this trend results from the need for the blend of business-to-business and consumer advertising skills we have already mentioned. Although numerous specific reasons for this evolution might be articulated, the answer seems to be the high-tech firm's desire for an agency with a new perspective and willingness to go beyond the traditional industrial approach. If one accepts the premise that high-technology marketers and markets are unique, then the rationale for specialization need not be further elaborated. The development of one high-technology advertising agency is illustrated in the following vignette.

The Quintessence of a High-Tech Advertising Agency

Reiser Williams DeYong (RWD) of Irvine, California, typifies (and is a pioneer in) the new breed of advertising agency; founded less than ten years ago (1975), RWD's approach reflects the view that the high-tech company's customer (business or consumer) is sophisticated and complex. This same high-tech customer has extensive experience with well-developed consumer advertising. The agency considers itself "the first consumer-tech agency."

Richard J. Reiser, the current president and RWD founder, brings to his high-tech advertising agency the same enthusiasm for creativity that generally has become expected in consumer advertising, but is more of a rarity in business-to-business advertising. Dick Reiser says, "We (felt) that engineers were also people with emotions like everyone else, who responded to automotive and beer commercials like the vast majority of the consumer audience."[a] With this philosophy firmly operational, RWD gave a personality to a computer terminal (creating the term *dumb terminal*) and has employed consumer media for effectively positioning its clients' products. Mr. Reiser has been actively involved in raising the communications consciousness of high-tech marketers, particularly in the computer field, by his chiding for better marketing planning and marketing research. He recently told attendees at a microcomputer conference in San Francisco in May 1983, that "advertising (now) will require levels of expenditures that most of you couldn't have imagined in the past. . ."; considerably more than the historic 1 to 3 percent of sales that is an outgrowth of its industrial advertising perspective. RWD's success is reflected in its $32 million (and growing) billings.

As of July 1, 1983, RWD became a wholly owned subsidiary of Cunningham and Walsh, a major U.S consumer agency, but will continue to reflect its high-tech special qualities. According to Mr. Reiser, his agency simply needed the type of help in research and media planning that a larger company, a Cunningham and Walsh, could provide. He believes that this move parallels the action of the smaller high-tech manufacturer that seems inevitably to merge with one of the giants in its industry. In other words, the sort of industrial evolution described in chapter 3.

An example of the highly creative advertising that RWD produces for CXC Corporation and its other clients is illustrated in figure 6-5. Among RWD's current clients are NCR (microcomputers); Tandon Corporation (disc drives); National Advanced Systems (computers); State of the Art (business software); Eaton Semiconductor; Advanced Electronic Design; and Gavelin (portable computers)

[a]This vignette is based in part of Chuck Wingis, "A New Breed of Ad Agency," *Advertising Age,* June 20, 1983, pp. M-10, M-11, and on personal conversations with Richard J. Reiser.

Note: Reprinted with permission of Reiser Williams DeYong, Inc., and CXC Corporation.

Figure 6–5. CXC Corporation Advertisement

Sales Promotion and Public Relations

Why are sales promotion and, particularly, public relations such important components of the high-technology firm's communication? Because of the nature of the high-technology presentation, sales promotion materials (brochures, catalogs, videotapes, etc.) must carry a major share of the hard information load. Although this role is discussed more fully in the personal selling section in chapter 2, it is important to note that the message to be presented via media must be carefully coordinated with the decisions regarding the sales promotion materials. The advertising manager needs to be responsible for both activities, or at least heavily involved in the coordinating effort. Further, all sales promotion efforts need to be handled in a way that is consistent with the company's desired reputation. One writer has said that promotion materials can either impress or inform.[28] For high-tech products and processes, our goal would be both.

Public relations also plays an extremely important role for the high-technology producer—a role that is usually more proactive than for the typical consumer or industrial products company, because so many of the public relations announcements the high-tech company wishes to make are actually newsworthy. A Ford or GM or P&G may go to great lengths to try to gain some publicity (business news space) for its new model or new product, but often falls short of its goals because some hard news occurs that day. By contrast, new products or processes in the biomedical, robotics, laser, and fiber optics fields intrinsically have reader interest. Accordingly, many high-technology companies have learned to coincide the timing of their announcement of new breakthroughs with some other event, perhaps a trade show in which they are participating, to gain maximum impact with their target market. So, corporate advertising and public relations goals need to be complementary and the messages consistent. Given this potential for the use of public relations in a planned way, the high-technology firm finds itself with another potent weapon in its communications arsenal.

Notes

1. Philip Maher, "Coming to Grips with the Robot Market," *Industrial Marketing,* January 1982, p. 93.

2. Roger C. Bennett and Robert G. Cooper, "The Misuse of Marketing," *Business Horizons,* November-December 1981, p. 54.

3. Maher, p. 94.

4. 1982 Annual Reports.

5. Edward G. Michaels, "Marketing Muscle: Who Needs It?" *Business Horizons,* May-June 1982, p. 66.

6. Philip Maher, "R_2D_2: Not Your Typical Robot," *Industrial Marketing,* January 1982, p. 94.

7. Robert A. Kriegel, "Anatomy of a Marketing Communications Plan," *Business Marketing,* July 1983, p. 72.

8. Richard T. Hise, Peter L. Gillett, and John K. Ryans, Jr., *Basic Marketing: Concepts and Decisions* (Cambridge, Mass.: Winthrop Publishers, Inc., 1979), p. 270.

9. Hise, et al., p. 270.

10. Kriegel, p. 72.

11. Dean M. Peebles and John K. Ryans, Jr., *The Management of International Advertising* (Boston, Mass.: Allyn & Bacon, Inc., 1983), p. 25.

12. Kenneth F. Englade, "Simply Understanding High Tech," *Advertising Age,* June, 20, 1983, p. M-12.

13. Peebles, p. 25.

14. Chuck Wingis, "A New Breed of Ad Agency," *Advertising Age,* June 20, 1982, p. M-10.

15. Glenn V. Ostle and John K. Ryans, Jr., "Techniques for Measuring Advertising Effectiveness," *Journal of Advertising Research,* June 1981, p. 20.

16. Peebles, p. 219.

17. Peebles, p. 224.

18. Kriegel, p. 74.

19. "Wired for a 'Word from Our Sponsors'," *Advertising Age,* June 13, 1983, p. M-28.

20. Adair Cunningham, "Cableshop Scores with the Longer Ad Pitch," *Advertising Age,* June 13, 1983, p. M-36.

21. Kenneth White, "Advertisers Make Connections on the Videopath," *Advertising Age,* June 13, 1983, p. M-42.

22. Uday Gupta, "Teleport Plan for N.Y. Area Rings Up Trouble for Cable TV," *Electronic Media,* June 9, 1983, p. 10.

23. Murray Roman and Bob Donath, "What's Really Happening in Business/Industrial Telemarketing," *Business Marketing,* April 1983, p. 82.

24. Tamar Lewin, "The Patent Race in Gene-Splicing," *New York Times,* August 29, 1982, p. F-4.

25. Philip Maher, "Hybrid Corporate Advertising Skirts the Budget Ax," *Industrial Marketing,* December 1982, p. 41.

26. "The 15% Media Commission is on the Way Toward Becoming the Relic in Ad Agency Compensation Plans," *Marketing News,* June 10, 1983, p. 9.

27. Peebles, pp. 117–118.

28. George Weingarten, "How to Manage Collateral with the 'Total Literature Concept'," *Industrial Marketing,* September 1982, p. 80.

7 Processing and Pricing

Two areas of marketing strategy—product planning and presentation (advertising and promotion)—have received our primary focus up to this point. Does this mean that other tools, such as distribution channels and pricing, are of lesser consequence to the high-technology marketer?

There does appear to be some deemphasis of the role and importance of distribution channels and pricing today. Companies seem to have less control over their distribution and pricing decisions, due to competitive and legal factors. This down-playing can be attributed partly to the perception that product planning and presentation have more glamour.

Consider the following recent observations regarding distribution channels and pricing:

> In evaluating the shakeout in personal computers, *Business Week* indicated that the successful firms would be those who master three areas; these included distribution and low-cost production or pricing.[1]
>
> A microcomputer industry spokesman said that in the microcomputer industry (hardware/software) "the success of an enterprise is directly related to the effectiveness of its distribution system."[2]
>
> A leading patent attorney suggested that the firm's pricing structure is one of three strong marketing determinants that "vitally affect an invention's commercial fate."[3]

Other comments in this vein have been made about high-technology industries and, certainly, distribution channels and pricing are to be regarded as critically important elements in most high-tech firms' marketing mixes.

In this chapter, distribution channels, or processing as we call it, are examined first. We consider the more traditional ways of conducting consumer and industrial distribution and then note the different dimensions that are unique to high-tech companies. Next, we discuss the types of channel relationships that were identified by our studies of high-technology firms. We then look at the role of the high-technology salesperson and at the special emphasis given to applications identification in fields such as robotics. Later, the approaches to pricing that are employed in marketing high-technology products and processes and the difficulties associated with pricing are considered.

Basic Distribution Channels

Philip Kotler has suggested two notable reasons for the importance of channel decisions: (1) "the channels chosen for the company's products ultimately affect every other marketing decision" and (2) "they involve the firm in relatively long-term commitments to other firms."[4]

Decisions regarding all elements of the marketing mix involve trade-offs. For instance, because the use of company salespersons and advertising both involve communicating with customers and potential customers, there may be certain trade-offs possible between these two marketing elements. As advertising usually offers a much lower cost-per-contact than does personal selling, a company may attempt to shift a greater portion of the selling job to advertising in attempting to reduce its current operating costs. Or the use of industrial distributors or representatives for a certain portion of a company's market is often more cost-efficient than adding company salespeople or starting a sales force, especially if the market has not been fully developed or the company has a very limited line of products.

Once the high-technology firm's channel arrangements are in place, its ability to make trade-offs (i.e., its flexibility) will be reduced sharply because the channel arrangements generally involve mid- to long-term contractual agreements. Fortunately, other marketing mix items, mainly promotion and price variation, retain much of their short-run flexibility, and therefore, from time to time, may need to be adapted to complement the channel arrangement.

Distribution Channels Defined

Stated concisely, the term *distribution channels* refers to the system the company develops for selling and moving its products or processes to its customers. The approach can be very direct: the company may employ its own sales force, handle its own credit arrangements, and ship its own goods, or it can be quite complex, involving a number of intermediaries. Perhaps the most common channels mistake is to consider only the transactions involved in this system when making decisions and to ignore the need to move the products physically and to handle credit, insurance, etc. These latter items become important considerations in any channel decision.

High-tech companies that are engaged in marketing consumer products find their channel alternatives very similar to those used by any consumer goods manufacturer. Therefore, their channel may involve wholesale and retail establishments and look something like this:

Producer ⟶ Wholesaler ⟶ Retailer ⟶ Consumer

Under this very traditional arrangement, the wholesaler would take title to the product, store it, and handle its distribution. However, if an agent middleman were used instead of a wholesaler, the channel would have the same number of stages, but the producer or manufacturer would retain title, physical control, etc. Obviously, this would increase the company's costs (credit, inventory, and the like) and delay the receipt of its proceeds. But, retaining title would give it greater control over its product's sale and possibly increase its profits. All of the considerations relate to the high-tech consumer product company's decision on what type of channel to use and suggest that all the alternatives have operating cost and potential profit plusses and minuses.

Basically, the industrial or business-to-business high-tech supplier has similar channel decisions to make, although the channel members may carry different titles. The high-tech business-to-business company may sell directly to all or a few of its customers, or it may move its products or processes to market via a network of intermediaries. The latter may include specialists in selling, credit, storage, shipping, international documentation, insurance, etc., or it might employ a wholesaler who takes title and possession of the product or process and from thereon handles all further marketing actions. While cost factors definitely influence channel choice, there are certainly other important considerations for the business-to-business marketer. These include:

customer needs;

market coverage;

effective sales efforts;

delivery time;

company reputation; and

inventory levels.

For example, a company may prefer to employ its own sales force or use sales agents (outside organizations selling on commission) in areas of the country or world where maintaining control over marketing operations is believed to be essential, while employing wholesalers in less important markets. This diversity can be illustrated by the multinational firm that uses its own sales force in the United States, Western Europe, and Canada, but sells through sales representatives or wholesalers in the Far East, South America, and the Middle East.

The various traditional consumer and industrial channels are shown in figure 7–1. Each, of course, has its own peculiar advantages and disadvantages, and many firms, therefore, use combinations of these alternatives. This is illustrated by well-known consumer food product companies which may employ their own sales force to reach the large food chains, agent middlemen to contact the remaining food retailers, and a different sales force and middleman combination to reach their institutional market (i.e., military, hotels/motels, or hospitals). The use of combinations is often employed when a company, such as a Smuckers or a Sara Lee, has both consumer and industrial/institutional markets for the same basic product lines.

How important a factor is direct marketing? Direct marketing involves the use of mail, the telephone (telemarketing), or even television to obtain sales and/or leads. Industrial companies have tended to employ telemarketing for low-volume accounts—those customers or prospects whose level of purchases make it economically unsound to have salespeople visit them regularly. (Often a combination of occasional visits and follow-up telephone calls are used.) A study reported in the August 1982 issue of *Industrial Marketing* (now *Business Marketing*) revealed that nearly three-fourths of the sample industrial firms employ direct response marketing for at least some of their products.[5] This finding is not surprising, and we would expect the totals to grow as the costs of personal selling rise (the estimated average cost of an industrial field sales call in 1981 was $178).[6] In fact, a report prepared for the National Association of Wholesalers-Distributors entitled "Future Trends in Wholesale Distribution" indicates that "inside-telephone salespeople will comprise 50 percent of the average wholesale distributor's sales [force by 1990], compared with less than 30 percent in 1980."[7] In addition, one can expect direct marketing to play an even greater role in the sale of higher-priced products and in reaching the better accounts and prospects. Again, this is partially due to the combination of increased costs and improved technology, but also to a recognition that direct marketing has many plusses to offer in obtaining and qualifying leads.

To summarize, we have attempted to review broadly the traditional channels employed in consumer and business-to-business marketing. We recognize that the whole idea of the distribution channel is new to some readers. What is critical to remember is that the industrial or consumer marketer is attempting to reach a target market by the optimum means possible. In order to do so, the marketer must evaluate customer needs and desires, relative costs, the desirability of maintaining some flexibility, timing (inventory and transportation), coverage, level of sales effort, and the

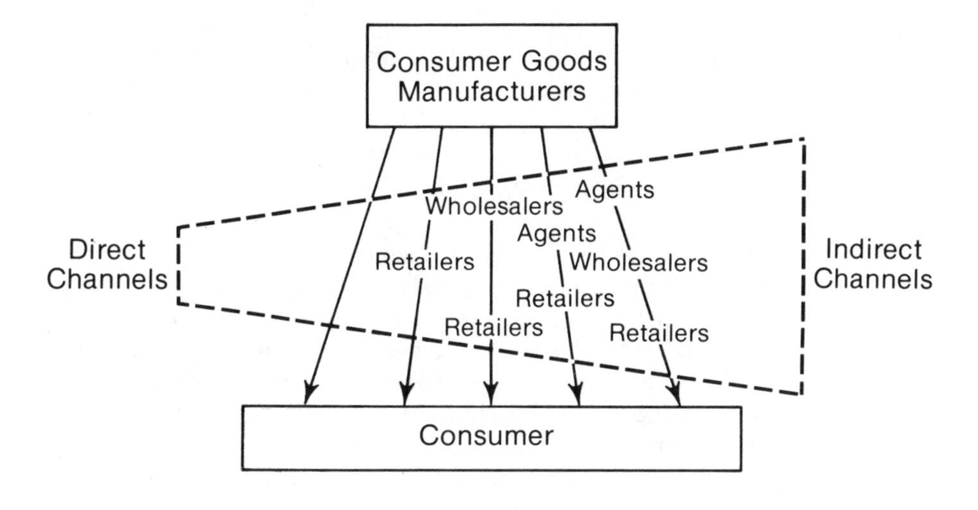

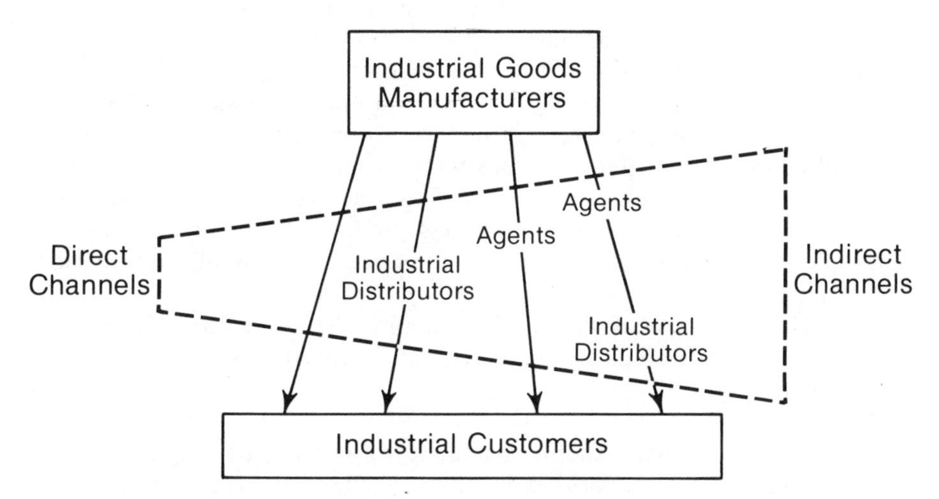

Figure 7–1. Traditional Consumer and Industrial Goods Channels

potential ROI or ROE under the various alternatives. The most attractive alternative may not be feasible (e.g., a desired agent or representative handles a competitor's account), or the company's line is too limited to employ its own sales force. Historically, we find many firms that used imaginative channels when faced with impediments and, in the end, fostered new and possibly even more effective approaches to reach their target market. Tupperware, Avon, Mary Kay, and L'Eggs are companies or brand names that come to mind when we think of creative consumer channels, and, certainly, Tandy's Radio Shack is illustrative of similar channel imagination.

High-Technology Channel Differences

The term *high-technology* has been treated almost generically to cover any product or process situation involving extensive R&D efforts. Yet, the differences between high-technology industries are perhaps no more apparent than when we consider distribution channels. At one extreme are the companies in the aerospace/aircraft industries that compete for government contracts or sales to the international aircraft markets, including U.S. airlines, and at the other would be the pharmaceutical firms that are heavily engaged in biotechnological research leading to a range of prescription products (consumer market) and products for research labs (business-to-business market).

Because of its product lines, Atari, a subsidiary of Warner Communications, Inc., competes in both the home video games and home computer market. So, at the retail level alone, it could be battling for shelf space in two areas of the same department store. Naturally, it would be selling in both toy stores and computer specialty shops—each with its own distinct channel possibilities. But, is it this wide variation in product lines and markets alone that distinguishes the high-tech channel from the channels of other consumer and business-to-business producers? Not really, because many non-high-technology companies produce extremely different products and services.

For a discussion of the channels employed by one major high-tech company, Corning Medical (a subsidiary of Corning Glass Works), see the vignette on page 157.

Is Fear the Difference? Another unique factor in the marketing of a high-tech product or process noted earlier is the gap between the technical developments and achievements produced by high-tech firms and the capacity of the marketplace to understand and utilize fully these developments. In fact, as we mentioned in the introduction, many of these developments have produced various kinds of fears among customers, potential customers, and the public at large. Would these concerns affect the nature of the channel and, thus, distinguish the high-technology channel from the more traditional ones?

There are basically two general types of fears or concerns at work; what we call the unknown and the "time warp." The first is the true fear of the unknown, which is especially apparent in those industries where the gap between technological developments and public or general understanding is the greatest. As Christopher Edwards, the editor of *Bio/Technology,* has said, "the common person currently brings to biotechnology a perception colored by science fiction fantasies"[8] and a host of other horrible visions, while even the more enlightened speak of dangers—often real—of nuclear spills, "herbicide clouds,"[9] and chemical waste sites. Although these may not affect directly the biotechnology producer's initial target market, say the breeder or the manufacturer, it could be of concern to the end product user—the ultimate consumer of beef or corn.[10] These fears might indicate the need for the company to employ its own sales force or to use agent middlemen or wholesalers who have the expertise and sensitivity to help cope with them.

Corning Medical's Distribution Channels

Corning Glass Works is involved in a variety of high-technology fields; for example, its ceramics group produced 70 percent of the outside fuselage surface (or windshield/tiles) for the space shuttle (see figure 7-2) and is responsible for many of the developments in fiber optics. In particular, Corning Health and Science Group, a division of Corning Glass Works, had total sales of $509 million (or roughly 32 percent of the parent's total) in 1982.

Look at how Corning Medical, one of the leaders in electronic/biotech diagnostic products used in hospital pathology and critical care departments, approaches its domestic customers. Basically, Corning Medical has four product lines: (1) pH/Blood Gas Analyzers, (2) Electrolyte Analyzers, (3) RIA, and (4) Electrophoresis. For domestic sales purposes, Corning Medical is organized on a product-line basis, each with its own approach to the market based on its customers' needs. For example, its Electrophoresis product line is sold through distributors, but Corning has its own technical support specialists to back up or service the customers. On the other hand, the company uses its own salespeople to sell RIA and instruments, but uses distributors to sell instrument consumables. Recently, Corning has begun to make extensive use of telephone marketing, with these sales handled by the customer service group.

Corning Medical's primary target markets, as mentioned earlier, are the pathology departments and the critical care areas (including respiratory therapy departments). Each department or area has its own primary decision-maker, the pathologist in the former and the respiratory therapy department in the latter. However, the selection of instruments or consumables is often influenced by others in the department. For example, in the pathology departments, the chief medical technologist and the medical technologist must also be considered by the channel members. (For more details on the structure of the sales organization, see pages 48–51.) While not directly part of the channel, Corning considers its field service organization, also organized along product lines, to play an integral role in its channel efforts by offering credibility to the company's emphasis on service. Corning has regional sales offices in Houston, San Francisco, Fairfax, Va., Chicago, and Medfield, Mass., and its regional sales managers oversee both direct sales and distributor activities within their geographic area.

CORNING

Windows. Protective coating. Retainers. These glass components cover *more than 70%* of the space shuttle's surface.

Each of them originates at Corning Glass Works. So do more than 60,000 other products, ranging from hair-thin optical waveguides to super-durable sidewalls for industrial furnaces.

Glass. Through Corning technology, it's versatile enough for the needs of earth – and tough enough for space.

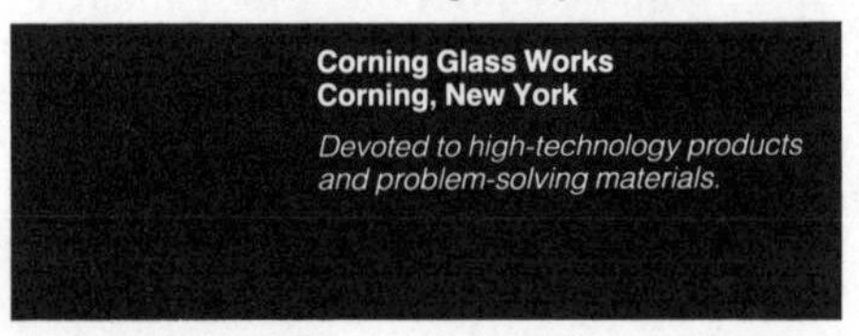

Note: Reprinted with permission of Corning Glass Works.

Figure 7-2. Corning Glass Works Advertisement

At a minimum, fear of the unknown forces marketing research to determine the level of product-specific concerns on the part of the target market and to suggest ways to combat them. One need only be cognizant of the magnitude of the public outcry and the price the company has paid in the ten-year-old Nestle infant formula or breast-milk-substitute controversy for its "initial misunderstanding of the nature of the issue. . .," to appreciate this point. (As a recent Nestle publication says, the price the company paid was not in economic terms "but in terms of harm to its public image and, more important, the consequent threat that the company might be prevented from meeting the real needs of Third World infants."[11]) In the biotech field and, particularly, genetic research, there are additional concerns that relate to religious and ethical issues. The depth and nature of these issues should also be probed as an indicator of overall market strategy and resultant channel problems.

The second fear is equally real and clearly effects on high-technology channel decisions as well. It is the anxiety of customers and middlemen alike that they will get caught in a state-of-the-art technology "time warp," that is, that their purchase (customer) or inventory (middlemen) will be outdated by new developments and they will be left a generation behind. This concern has been evident particularly in the computer (and computer peripherals) and office automation areas. As a senior vice-president for bank operations administration at AmeriTrust (Cleveland) recently stated, "one of our fears is that [our equipment] will become obsolete . . ." and thus, they are "moving cautiously [on such purchases]. . . ."[12] Then, there is the closely related misgiving that there will soon be a shakeout in such high-tech fields as the personal or microcomputer, software, and robotics industries. In other words, there will be a glut of producers and the markets will not be able to support all of them. In robotics, one industry executive sees a shakeout coming "even sooner than people have been predicting . . .";[13] and this circumstance may not be surprising, considering that there are some two hundred Japanese and one hundred U.S. firms, plus a scattering from other countries, in the field.[14] And it has been estimated that only a fraction of the 150 or so microcomputer manufacturers active in 1983 will still be competing in 1987.[15] Thus, being left with obsolete equipment or equipment produced by a company that has left the industry is another dread shared by customers and middlemen. Concerns, of course, felt by a company's middlemen can have an enormous negative impact on its channel relations.

Is Overcrowding the Difference? Finding the optimum channel is at best a difficult undertaking, especially for a new company or new entrant in an industry. Add to this a situation in which a field is overcrowded and most firms will feel uncertain about the best route to the market. This kind of a problem has been reported in the personal or microcomputer area and its related businesses. Philip Maher, a *Business Marketing* senior editor, has called Comdex (the annual industry trade show in Las Vegas) the "largest mating ritual in the business world," as the microcomputer companies seek the "perfect distribution channel." So many potential corporate buyers of the micros have not adopted coherent buying patterns that the alternatives include direct sales, dealers, distributors, OEMs, wholesalers, retail stores, company-owned stores, system vendors, etc.[16] And because many microcomputer manufacturers use multiple channels, the result has been pricing confusion, dealer discontent, poor customer service, and similar concerns. It is not unexpected that this high-technology field is considered ripe for a shake-out and that the competitive advantage often lies with the major MNC with the established reputation.

What Do Our Study Findings Imply? While our inquiries covered a number of firms in a variety of high-tech industries, our channel-specific survey focused on the robotics field. Firms in this industry are still in an early stage of the s-curve (chapter 5), and the industry itself has many features inherent in supply-side marketing—it is R&D driven. Since so many high-technology industries have these characteristics, the robotics field was believed to be an appropriate selection for our distribution channel study (see vignette titled "Robotics Industry Channels," p. 161).

Most of the companies in our robotics industry sample employ multiple channels to reach their target markets. Over 90 percent of the robotics companies have their own sales force, yet, interestingly enough, none relied solely on its sales force to represent the company in the actual selling effort. Heavily involved in the channel activities are the application support department, industrial distributors or agents (reps), and even, on occasion, the R&D department.

After the potential opportunity has been identified, the salesperson typically handles the initial contacts with the prospects. However, in about two-thirds of the companies, the robotics engineers/technical people become actively involved in the later phases of the sales process, where specific applications or designs become critical. (In most of the companies, the salespeople have engineering backgrounds).

To sum up, the firms in the robotics field rely on a variety of sources including field sales personnel, industrial distributors, manufacturer's reps, and agents to initially identify prospects. Once identified, typically the company's own sales force takes care of much of the early contact work, but is subsequently backed up by the applications support staff or other technical personnel in the later stages of the sale. For our purposes, the fact that slightly more than three-fourths of the robotics firms (78 percent) use middlemen (manufacturer's reps/agents or industrial distributors) is sig-

Robotics Industry Channels

During March and April of 1983, the authors conducted an extensive survey to determine the distribution channels being employed by firms in the robotics industry. First, the results indicated that most firms use middlemen (industrial distributors, manufacturer's reps, etc.) and their own sales force, as well as an applications support department, to represent them in their selling effort. Percentages of usage were:

Applications support department	77%
Company sales force	96%
Middlemen	81%
R&D	19%

In other words, the findings show that the bulk of the robotics companies use various approaches (and combinations) as they move from merely identifying prospects to completing the sales. Second, some 92 percent of the firms indicated that their sales force was one of their primary sources for identifying prospects, while one-half included referrals by middlemen as one of the primary ways they identified prospects. Third, roughly half (54 percent) of the respondents said that they make rather extensive use of middlemen to handle at least a portion of their selling and marketing activities (i.e., middlemen were typically used on a regional basis). Among the middlemen, the most often employed were manufacturer's representatives.

When robotics firms do use middlemen, the nonselling services that are provided range from applications support to financing. The percentage of usage of each service by the robotics firms is presented below; applications support is clearly the service used most often.

Services Provided by Middlemen
(percentages based on usage)

Service	*Extent Used*
Applications Support	86%
Installation	35%
Storage	15%
Transportation	12%
Financing	15%
Advertising and Brochures	23%
Customer Identification	65%

nificant. The combination of a company's own sales force and either industrial distributors or manufacturer's reps/agents appears to be quite characteristic of high-technology companies.

Have We Answered the Question? Is the high-technology firm really different in terms of distribution channels, and if so, how? It would be easy to argue that Texas Instruments or Digital Equipment Corporation are not greatly different from other major companies that market both to individual consumers and business. Perhaps they are competing in marketplaces characterized by uniquely heavy competition and rapid state-of-the-art change, but they still face the age-old struggles for shelf space and buyer attention. And, of course, many high-tech firms are not focusing on dual markets.

Still, it is apparent that high-technology industries or firms typically are faced with inordinate customer and public fears; the need for a supply-side approach (high reliance on identifying applications); confused paths to their markets; both highly sophisticated and highly unsophisticated decision-makers in their target markets; and other peculiarities. The same considerations noted earlier in this chapter regarding the selection of a distribution channel are appropriate for the high-tech firm. In addition, the high-technology firm must select a channel(s) that is sensitive to its particular customers' or publics' fears and that recognizes new and often strikingly different applications. We advise that the heavy involvement of the firm itself is essential, even when middlemen are involved, during the early stages of the market's development (supply-side phase). And the high-tech company must constantly be sensitive to middlemen's concerns (channel conflicts) whenever dual channels are employed.

Avoiding Channel Conflict

The whole area of the use of the company's own sales force was discussed in chapter 2. As we indicated there, one unique characteristic of high-technology firms is that most employ some form of sales organization, even when they rely on outside intermediaries to handle the bulk of their selling activities. Perhaps they use only their own salespersons to handle *Fortune* 1000 firms, which is Apple Computer's approach,[17] or to key on a geographic area or customer group, but they still have some portion of their accounts handled by middlemen.

At the other extreme, of course, are the manufacturers who employ their own sales organization for 90 percent or more of their sales and simply use some type of middlemen for their very small customers. Then, we have the type of channel mix that is now found in the microcomputer and software fields where companies may be using one complex set of channel arrangements to sell personal computers and software to individual households and another to sell to smaller businesses. For example, what happens when a businessperson visits a computer retail store and makes a purchase for joint home and office use? As might be expected, this channel mix can create considerable redundancy of effort, confusion among possible customers, and channel conflict situations.

Another special case is offered in the robotics field where the company's sales force and applications engineers may work together with middlemen (manufacturer's reps) on a possible account. Or, in fact, either (salesperson or manufacturer's rep) might call on the other for assistance and both will turn to the same applications department and service group at the home office for direction and follow-up efforts. Since a big portion of the sales training in the robotics fields, as well as other high-technology fields, involves learning about the company's products and is generally technical in nature, it may well be that both the company's new salespeople and the manufacturer's reps' salespeople find themselves being trained together. (A discussion of manufacturer's reps is presented in the vignette on page 164.)

What are the typical kinds of channel conflicts that may result from mixed channel arrangements and potential overlap of sales responsibilities? In the very small organization, the kinds of mixed approaches we have described may create few problems, as long as there is mutual respect and what might be termed *share objectives.* A company may have only a few salespeople covering enormous territories and reaching only large companies, for example, the eastern half and the western half of the United States, and in addition, sell through a couple of wholesalers or sales agents to everyone else. On occasion, each may stray into a gray area of coverage, but if the issue is handled promptly and equitably by the home office, conflict may be avoided.

If a company has salespeople who along with wholesalers and reps handle business-to-business contacts, and its own sales force plus wholesalers and retailers to reach the consumer market, the chances for conflict between the two sides is greatly increased. To avoid conflict, the marketer needs to recognize what the likely sources of conflict are and take advance precautions to avoid them. Some possible conflict areas, of course, such as delivery time, inventory level requirements, adequacy of margins, and others, are found in all channel arrangements. However, unique to the types of complex or mixed channels found in many high-technology industries would be conflict resulting from the following situations:

the intermediaries perceive that the company reserves all the better accounts for its own sales force;

the company has not identified and handled by policy all the various possibilities of territory or customer-type overlap, (i.e., between the sales force and intermediaries or between various intermediaries);

the intermediaries perceive that the company's sales force is given preference when disputes arise; and there are disputed sales and the company has not clarified the methods of determining who should receive credit for them.

This list is by no means exhaustive, but it does signal the kinds of possible problems that should be anticipated and resolved in advance. Typically, they relate to territory or customer definition, determination of credit/ compensation, and matters of equity. The charge of favoritism toward the company's own sales force or retail outlets is common from outsiders and often requires continual monitoring to ensure a high level of middleman cooperation.

Are Reps Right for You?

Selling through manufacturers' reps is not the way to go for all firms. In fact, some firms require a direct sales force under tight management control. For others, however, selling through reps is just what the doctor ordered.

New, growing firms with limited resources can profitably bring their products to market through reps because of the contingent nature of a rep's compensation. Large sales of commission expenses are not incurred until after a sale is made.

Established firms with a single product or product line outside the mainstream of their normal business can profitably use reps to sell the exceptional product or product line.

Established firms whose products are sold in conjunction with the products of others in a package or a system can profitably use reps to bring those products to market.

Firms feeling the pinch of the recession, and with territories that cannot support salaried salespeople, might profitably use reps in the weak territories.

Companies supplying markets traditionally served by reps or sales agents can benefit from using those customary channels to reach their customers.

Under conditions like those, reps offer a number of advantages which are not available through salaried or direct selling operations.

Lower Fixed Selling Costs

Reps do cost less than salaried sales people, often considerably less. Yet even through reps are not paid until after the sale has been made, they are by no means free until then. They still require guidance, product literature and training. Furthermore, because sales reps are somewhat like volunteers, free to go at their own option, they may require higher quality service and support than a salaried sales force would tolerate.

Better Knowledge of Their Markets

One thing an established rep should be able to offer a principal is a thorough knowledge of the accounts that make up his market. Because reps carry several related and complementary lines, they should know all or nearly all of the potential users of any of them. The knowledge a rep has of his territory can be much greater than that of a salaried salesperson carrying the line of a single manufacturer.

More Frequent Calls and Better Account Probes

Because a rep carries several lines, he has more reasons to call on an account and tends to make more frequent calls than a salaried salesperson. He should, therefore, be more attuned to the needs of an account and better able to spot an opportunity.

Minimal Sales Management

A salaried sales force requires the daily or weekly monitoring of the number, the allocation and the quality of its sales force. When a rep is engaged, he provides those essential sales management functions. The principal is not required to perform them.

Independent Status

Because a rep represents several principals and not just one, he tends to be viewed by his accounts as an independent, honest broker. He lends his personal authority and reputation to the lines he carries. His accounts assume that he will jeopardize neither reputation nor authority by carrying an inferior line or representing a second-rate principal.

Stewart A. Washburn, "Are Reps Right for You?" *Business Marketing,* June 1983, p. 86. Reprinted with permission.

One of the biggest concerns relates to the firm's pricing policies regarding the different avenues of distribution. It must be aware that the use of overlapping channels can result in channel price competition, as the various channels tend to undersell each other.

Pricing

Marketers must learn to deal with the many gray areas of decision making; there are few instances in which we have enough information to allow one to make simple black or white decisions. Nowhere is this blurring more evident or our techniques more inadequate than in the area of pricing and pricing strategy.

How important is pricing as a tool or tactic of the high-technology marketer? Many consumer and industrial good producers would answer that pricing is essential, because their research or intuition has indicated that the sale of their product or service is highly sensitive to price changes (i.e., a slight price rise decreases units sold and a slight lowering of price produces increased sales). As the microeconomist would say, they have found that they have a high price elasticity for their offerings. The marketer of high-technology products, however, may find that he or she has a much different pricing situation, one that necessitates a judgmental decision rather than a data-based decision. Some businesses, of course, prefer to compete via non-price competition,[18] but in this instance, judgmental pricing is due to the pricing complexity one finds in rapidly evolving industries. There is simply no historic pattern to offer needed data. In this section, we will review the role and problems of pricing in high-technology industries; look at some basic pricing alternatives; and indicate the factors to consider in developing a pricing policy.

Purpose of Pricing

Does the price of a product or process indicate more to prospective or current customers than merely "the amount necessary to complete a transaction?" The answer is yes, since we have come to expect price to offer us a range of signals about an industrial or consumer product.

For example, the knowledgeable buyer uses price as an indicator of:

1. The level at which the company competes;
2. The relative value between competing products;
3. The image the company wishes to project (i.e., leader, follower, etc.);
4. The company's state-of-the-art position; and
5. Most importantly, relative quality.

One may argue that the above are not mutually exclusive, and this is correct. However, consider the first indicator—the level at which the company competes. Automobile consumers recognize that the Pontiac Firebird competes with the Toyota Celica GT and, in fact, automobile companies often identify the models with which they compete by comparisons provided in their literature or advertising. Other companies in high-tech industries do this as well. For example, through an advertisement Commodore Computer told the *Business Week* reader that Commodore's competition is Apple, Tandy (Radio Shack), and IBM[19] and provided the prices of the competitors' equipment along with its own. Commodore's ad indicated the competitive level at which the company wished to be perceived. Often the signal as to level is not this direct. When a robotics firm (International Robomation/Intelligence) passed out literature about its $9,800 robot at an international robotics show in Chicago in 1983, it was in effect suggesting its competitive level and providing several other price signals as well.

Of special importance for the high-tech firm is the need to have a pricing policy that is consistent with its desired image. It is rarely credible to suggest that a company has the highest quality in an industry and also the budget price.

Pricing Problems

When is pricing a very difficult problem? Speaking of pricing in a broader context than high-technology marketing, Philip Kotler has pinpointed four difficult pricing situations:

1. when a firm must set a price for the first time;
2. when circumstances lead a firm to consider initiating a price change;
3. when competition initiates a price change; and
4. when the company produces several products that have interrelated demands and/or costs.[20]

The list is appropriate for the high-technology marketer, especially the first item. We need to add a fifth problem situation: when the high-tech firm moves from being a supply-side marketer to a demand-side marketer. This fifth problem situation would seem to be covered by the second situation listed above; however, number two simply refers to special situations, maybe periods of inflation or shortage where a price increase seems justified because of new cost/demand considerations or to situations where temporary price deals could stimulate demand. On the other hand, as the high-technology firm moves beyond the introductory phase, it faces a whole new marketing environment. Quite typically, for example, its product is less unique, its competition has rapidly increased, and its R&D lead has evaporated. Therefore, its pricing strategy needs to become more demand-centered (e.g., responsive to customer demands, target on niches, etc.) in order for it to survive.

Costs Predominate

There is a strong tendency for costs to play the dominant role in pricing, especially the pricing of new high-technology products. The reasons for this cost orientation are several:

1. Many high-technology companies either currently engage in, or have a history of engaging in government research contracts, which tend to be cost-plus agreements;
2. High-technology companies usually incur large development costs and often wish to recapture these costs as early as possible;
3. Smaller high-technology firms often have limited experience in pricing a product or process and a cost orientation seems to offer less risk; and
4. High-technology firms typically have rather solid cost data at their disposal, but their demand information, especially for new products, tends to be quite soft (e.g., rough forecasts, intuition, survey research).

The tendency to employ a cost orientation in pricing is certainly not limited to high-technology industries or to new products. A 1981 study of United States-based international industrial marketers indicated that costs, including transportation costs, production costs, tariffs and taxes, and other cost items, were believed to be more important factors than were "a competitor's prices" or "demand concerns" in overseas pricing.[21] A 1972 study of United States-based consumer marketers' international sales operations produced comparable results, for example, a strong cost orientation.[22]

In article after article on the developments within high-tech industries, one finds that the arrival of the large manufacturers—whether it be GE and IBM in robotics or IBM in microcomputers—is heralded as creating a new game in which the cost factors will shift in favor of these new players. It is understandable why this view might be held, because it is recognized that the inclination toward cost-plus pricing is widespread; that industry members may see that costs of competitive R&D, etc., will be increasing; and that the feeling persists that many high-technology industries will eventually be commodity producers. Clearly, it is based on the assumption that the lower cost producers—the ones with the economies of scale—will prevail even when lower cost is simply a relative term.

Economies of Scale and the Porter Curve. There certainly are cost-savings available to those firms that operate at the most efficient point in an optimum-sized production facility. There are also possible advantages available to companies that are able to spread certain marketing costs, such as for salespersons or advertising, over a wider product line, or have bargaining strength in financial circles, or are vertically integrated. However, economies of scale, despite the emphasis given to it in certain economics literature, does not necessarily determine who will remain when industry shakeouts occur.

The three generic competitive strategies that Michael E. Porter has suggested offer some interesting direction for the high-technology firm, especially those which see scale economies and a cost orientation as the only viable alternative to achieving survival and success. Porter indicates that *overall cost leadership* is one generic competitive strategy, and it does involve "efficient-scale facilities, vigorous pursuit of cost reductions from experience, tight cost and overhead control, . . . and cost minimization in areas like R&D, service, sales force, advertising. . . ."[23] Porter contrasts two other generic strategies that create positions that can be defended in the long run and can cause companies to outperform their industry competitors. The first is what Porter calls *differentiation,* and it requires a firm to create something that "is perceived industry-wide as being unique."[24] Corning Medical attributes much of its success to its service efforts. That is illustrative of being unique. Other ways firms have of differentiating themselves include: design or brand image, technology, and dealer network. It is this act of differentiation that can allow a firm to succeed and still not be the low-cost producer. In fact, a quality image or a strong dealer or distributor network may even be a greater entry barrier and afford greater protection from competition than, say, lower production costs.

Another alternative (generic competitive strategy) is what Porter refers to as *focus,* or what has popularly been referred to as targeting on a niche. In the chapter on positioning (chapter 3), we emphasized the importance to the high-technology producer—especially one developing a multiple-use breakthrough—of carefully identifying its correct target market(s). And we stressed the need for the high-technology firm to establish priorities in its markets or even industries. Under Porter's definition of focus, we see an even narrower position, that is, "a particular buyer group, segment of the product line, or geographic market . . . ,"[25] as an industry's needs become identified. (A number of stimulating illustrations of consumer product focusing are offered by Jack Trout and Al Ries in their book, *Positioning: The Battle for Your Mind,* describing various marketing strategies.)[26] Under Porter's third generic strategy, a firm's focusing allows it to achieve narrower but still viable cost economies or very specific product differentiation.

To clarify any questions regarding Porter's three generic strategies and to link them more tightly to pricing, look at figure 7–3. Porter employs a u-shaped curve to illustrate his three strategy concepts and to indicate how high return on investment positions can be achieved by firms with extremely different market shares. For discussion purposes, assume that figure 7–3 illustrates an entire high-technology industry with 24 or so competitors. One of the firms, we will assume, lies at the far end of the curve—at its highest

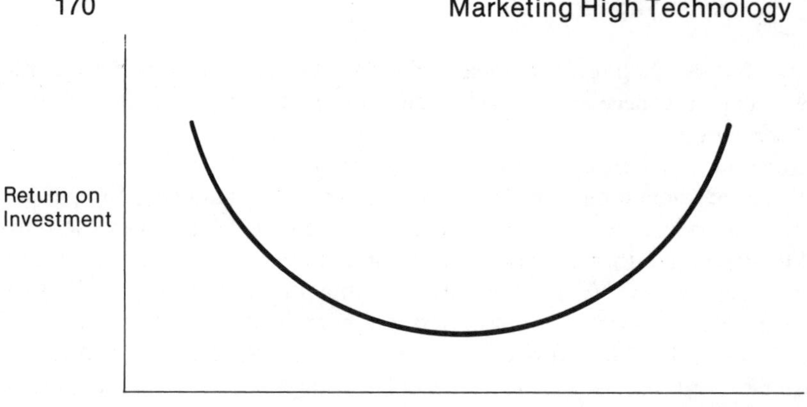

Note: Reprinted with permission of Macmillan, Inc., from *Competitive Strategy: Techniques for Analyzing Industries and Competitors* by Michael E. Porter. Copyright © 1980 by The Free Press, a Division of Macmillan Publishing Co., Inc.

Figure 7–3. The Porter Curve

point—indicating both high share of market and high ROI. We will further assume that it has achieved this strong position as a result of a differentiation strategy (dealer network and positive image)—one of Porter's generic strategies.

At the same time, a second firm is focusing on a specialized niche in the market. While its overall share of market is low, within a very narrow submarket it is dominant. Thus, it could be located at (or near) the highest point on the left side (or end) of the Porter Curve and, consequently, be producing an ROI comparable to the first company. In effect, this second firm is also employing a generic strategy of Porter's—it is a focusing strategy.

But what of the other 22 firms in our hypothetical industry? Let us assume these companies are not taking any particular differentiation action or implementing the other strategies we have discussed. They are all aggressively competing for the remaining available market shares and have relatively low ROI positions. They are operating in the trough (figure 7–3) near the low point in the curve or, to rephrase their situation, these 22 companies are operating in a competitive box, whose parameters are established by the successful firms. They can be characterized as having:

no particular scale economies (i.e., production, promotion or finance);

no real distinction in terms of filling customer needs;

no major discernible or promotable differentiation, only superficial differences; and no effective focusing strategy (i.e., they have been unsuccessful in or have not attempted to narrow their markets by focusing on a particular niche).

These firms have become virtually generic as they try to cover all product lines and scramble to find something that will give them an overall state-of-the-art edge or some other unique quality. Historically, industries of this kind have been ripe for a shakeout—consolidations, acquisitions, or closures that leave a half-dozen or so firms in the industry. Headlines, such as "Chip Wars," vividly describe the competitive situations in the microchip industry, and similar battles can be seen in the robotics and software fields.

To return to pricing, contemplate what Ronald Gist referred to as perceived value. Gist used the simple equation

$$\frac{\text{Perceived Quality}}{\text{Perceived Price}} = \text{Perceived Value}$$

to illustrate that value can be increased by improving perceived quality just as readily as by decreasing prices. This conceptualization is especially significant for the high-tech firm that depends heavily on quality improvements.[27]

Pricing Objectives

Before concluding our discussion of pricing by looking at several specific pricing alternatives, let us briefly consider several realistic pricing objectives that high-technology firms may have. While we have alluded to most of these at some point in our discussion, they do bear specific attention. We must add that such objectives need to be consistent with the company's overall corporate and marketing objectives.

Maximum Profit. Microeconomists would probably consider profit maximization to be the most appropriate objective. Yet, difficulties in forecasting demand and in accurately allocating costs make the objective more appropriate in theory than in practice. More often, we see firms attempting to achieve a target ROI or a satisfactory profit situation.

However, as with other pricing objectives, companies need to decide whether they are seeking to reach their profit objectives over the long or short run. The time dimension selected is influenced by factors such as the patent situation or whatever the temporary advantage that affords the company a period of protection before the field becomes filled with competitors. Given the rather rapid product or process life cycle we are seeing in many industries today, firms appear to be thinking in shorter-range terms.

Inhibiting Competition. To acquire a bit more breathing space, a second pricing objective may be one that attempts to delay the rush of competitors to the industry. Often high-tech industries attract considerable attention simply because of the nature of their research and the glamour or publicity it generates (e.g., genetic engineering, software, or microcomputers). If firms in a new industry post extremely high profits, it even further accentuates the industry's perceived opportunity. Therefore, it may be to the existing firms' advantages to pursue a pricing approach that does not yield unusually high profits.

Pricing to Establish Relationships. In many high-tech industries, the initial sale has the potential to produce a lasting business relationship. For example, there are often engineering modifications, spare parts, accessory items, and consumable sales that result from the initial sale of a product or process, and they all may be tied to that first establishment of trust in the supplier. This marketing fact of life is true, for example, in the robotics field. If a positive association results from that initial sale, future product sales and a long working relationship often evolve. Consequently, achieving such a relationship could well be one prominent pricing objective of a firm.

Other Pricing Objectives. In setting their prices, high-technology companies may have several additional primary or secondary objectives. These would include: allowing for a satisfactory margin for their distributors; avoiding any unfavorable governmental action or general public reaction; maintaining the proper price gap with key competitors; or assuming a price leadership role. Each of these carries its own set of plusses or minuses, especially the latter two objectives. Both of these depend heavily on the actions of other firms and, therefore, are often less predictable and may result in more frequent price changes.

Back to the Basics

The most difficult pricing decision can occur at the time a new product or process is introduced. Later, so much depends on factors, for instance the speed with which competitors are entering the market or the rate of state-of-the-art developments, that the company may have less flexibility in its pricing approach. Therefore, the two pricing techniques that receive the greatest attention from high-tech marketers are the two fundamental new product or process pricing strategies—the skimming price approach and the penetration price approach.

As depicted in figure 7–4, these two strategies actually lie at the opposite ends of a pricing continuum, as skimming represents a high-price strategy and penetration a low-price strategy. While neither is frequently used to its extreme, each offers quite realistic approaches for the high-technology producer to follow, depending on its particular objectives, competitive situation, degree of protection, etc.

Skimming Pricing. The circumstances that permit the use of a skimming price approach are suggested in figure 7–4. Unlike the penetration approach, the skimming strategy can successfully be employed only if the firm has some particular state-of-the-art lead or form of competitive protection. Examples of consumer product companies which have successfully used skimming include Kodak, Polaroid, RCA (color TV), Sony, and a host of others. In fact, conducted in an optimum fashion, one would expect to see a firm initially employ skimming and then gradually move to lower prices as demand at each subsequent consumer price level becomes saturated. Through this careful orchestration of its pricing, the firm would simply be reducing its price as it moved through each subsequent stage in the product or process life cycle.

What has happened in so many high-tech industries is a collapsing of the protection time, which plays havoc with a carefully planned and organized pricing approach. Certainly, some companies do find themselves in positions to employ skimming pricing, especially when they have patent protection and when they are seeking to recapture R&D expenditures quickly or to improve their operating cost position. Even so, one is less likely to find anyone today in the enviable situation that Dr. Edwin Land was in as he methodically marketed the Polaroid camera.

Penetration Pricing. This strategy is typically followed for a new product that has little or no protection in an industry that is quite competitive. Here the lead time is extremely short and the objective is to achieve as large a market share as possible while the slight edge exists. A penetration approach might be quite appropriate in high-tech industries in which the state-of-the-art technology is virtually generic.

In contemplating the use of this approach, one of the objectives mentioned earlier needs to be kept in mind. If the firm's pricing objective is pricing to establish a relationship, a penetration pricing strategy might be most effective. By pricing the basic product low, the company could perhaps reap the benefits that are often byproducts of long-term supplier-customer relationships. Thus, the "$9,800 robot" might be a most effective pricing approach.

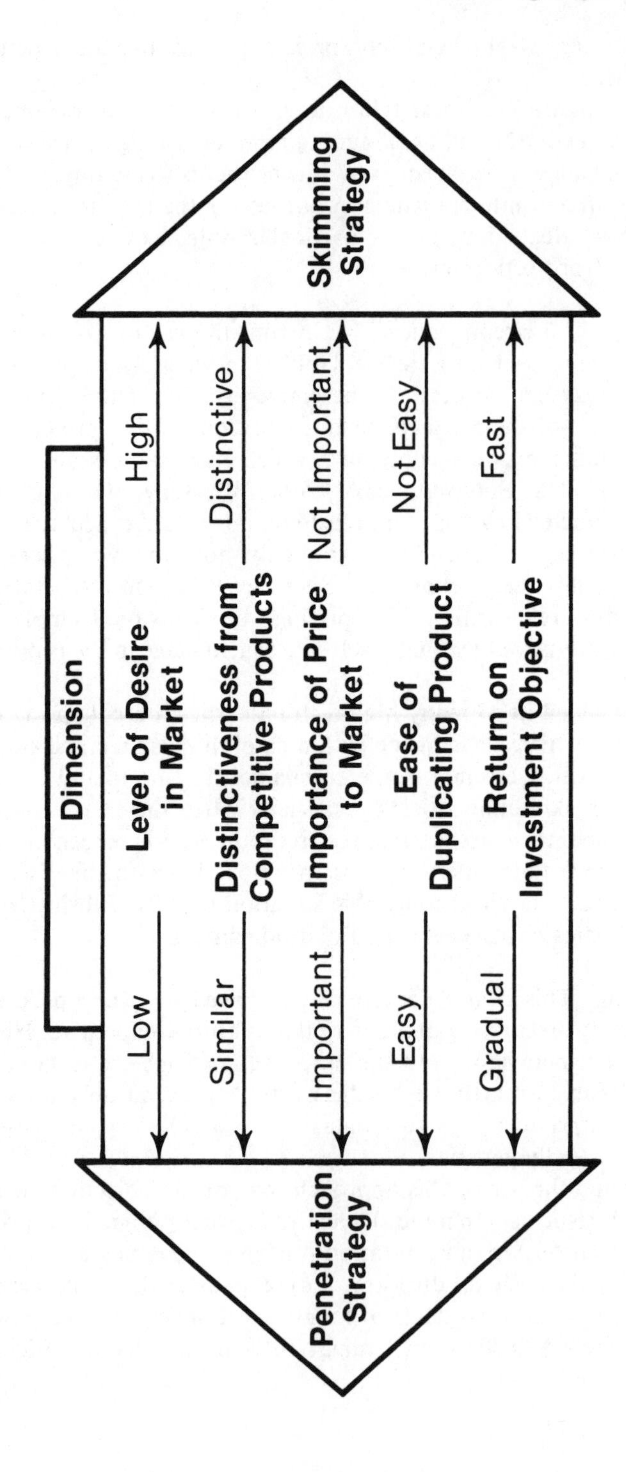

Source: Richard T. Hise, Peter L. Gillett, and John K. Ryans, Jr., *Basic Marketing: Concepts and Decisions* (Cambridge, Mass.: Winthrop Publishers, Inc., 1979), p. 450. The authors wish to thank Professors Hise and Gillett for permission to reprint this figure.

Figure 7–4. When to Use a Penetration or Skimming Strategy for Pricing New Products

Pricing Complexities in the Channel

While we have attempted to take a realistic view of pricing in this chapter, there are still many problems and complexities of pricing that we have not attempted to explore. Some are firm- or industry-specific, while others are too unique or too rare to make them appropriate for consideration here. And, of course, there are a few that simply require too much depth of explanation.

However, we do need to make mention of a few of the pricing dilemmas that do arise occasionally to plague the high-technology marketer, mainly problems involving sales through middlemen. These dilemmas include:

trying to develop a consistent pricing policy when marketing the same (or similar) product to different types of customers. This problem is illustrated today by the microcomputer field that has both home and small business application, yet the firm may have to consider retail store and wholesaler margins for one and direct sales commissions for the other;

trying to develop a consistent pricing policy when selling through different channels to reach the same customers. In the microcomputer field, vendors selling to business may go through direct sales, vendors, OEMs, etc. This can create pricing havoc "as competing channels undercut each other";[28]

trying to avoid channel conflict (or employee morale problems) when the company's own direct price to end users may be higher or lower than the price offered by distributors;

trying to reduce prices in a hotly competitive industry when distributors are reluctant or refuse to cooperate (lack of cooperation is frequently due to distributors' inventory, which may, in fact, have cost them more than the current end user price the company recommends); and

trying to determine the critical turns in industry demand so that pricing changes will keep pace.

Although the examples used come mostly from the microcomputer field, similar problems occur when competition becomes keen in any high-tech industry and when heavy reliance is made on distributors, wholesalers, and other middlemen. It is always tempting to vary price, that is, to be inconsistent in a pricing policy as sales become more difficult to make. A natural outgrowth of this inconsistency is "at a minimum" competitor retaliation and "at a maximum" channel chaos and industry pricing disarray. With these possible problems in the offing, is it not surprising that companies prefer nonprice competition?

Bid Pricing

No discussion of pricing would be complete without mentioning bid pricing, which plays such an important role in governmental contracts at all levels. Bid pricing is so specialized that we have chosen to segregate it. Unlike most of the pricing discussion in this chapter, bid pricing necessitates a cost-plus orientation.

Moreover, special expertise is required in all phases of the bidding process, for example, determining what contracts are outstanding, obtaining the appropriate procedural materials, developing the proposal, and submitting the bid. Earlier we discussed pricing to establish relationships and the points we emphasized are germane to bid pricing as well. Very often, the real profits in a government sale come from servicing, spare parts, etc., rather than from the new equipment or process itself. In addition, having a government contract often permits a vendor to hire experts in a given area or to develop technical expertise that it would otherwise be unable to do. These advantages need to be kept in mind by bidders as they seek to win contracts.

Notes

1. "The Coming Shakeout in Personal Computers," *Business Week,* November 22, 1982, p. 73.

2. Philip Maher, "High Tech Mating Rituals: Scoring the Right Distribution," *Business Marketing,* June 1983, p. 54.

3. David Pressman, "Predict Your Invention's Market Success with a Commercial Potential Checklist," *EDN,* February 17, 1982, p. 284.

4. Philip Kotler, *Marketing Management: Analysis, Planning and Control,* 3d ed. (Englewood Cliffs, N. J.: Prentice-Hall, Inc., 1976), p. 275.

5. "Industrial Marketing Survey: Business/Industrial Direct Response Promotion," *Industrial Marketing,* August 1982, p. 75.

6. Sara Delano, "Turning Sales Inside Out," *INC.,* August 1983, p. 99.

7. Delano, p. 99.

8. Christopher Edwards, "Consumers Must Be Prepared for Biotech Products," *Bio/Technology,* April 1983, p. 237.

9. "A Drug Giant Plagued by Dioxin's Poison," *Business Week,* May 2, 1983, p. 42.

10. Edwards, p. 75.

11. Maggie McComas, Geoffrey Fookes, and George Taucher, *The Dilemma of Third World Nutrition: Nestle and the Role of Infant Formula* (U.S.A.; Nestle S.A., 1983), p. 17.

12. "Computer Shock Hits the Office," *Business Week,* August 8, 1983, p. 47.

13. Richard A. Shaffer, "Market for Robots Turns Sour, May Speed Industry Shakeout," *Wall Street Journal,* April 22, 1983, p. 21.

14. "The Robots are Coming," *Barron's,* April 11, 1983, p. 8.

15. Maher, p. 54.

16. Maher, p. 54.

17. Maher, p. 54.

18. Ronald R. Gist, *Marketing and Society,* 2d ed. (Hinsdale, Ill.: The Dryden Press, 1974), p. 484.

19. Commodore Business Machine advertisement, *Business Week,* May 23, 1983, p. 163.

20. Philip Kotler, *Marketing Management,* 3d ed. (Englewood Cliffs, N. J.: Prentice-Hall, Inc., 1976), pp. 249–250.

21. James C. Baker and John K. Ryans, Jr., "International Pricing Policies and Practices of Industrial Product Manufacturers," *Journal of International Marketing,* Vol. 1, Number 3, 1982, pp. 127–133.

22. J.C. Baker and J.K. Ryans, Jr., "Some Aspects of International Pricing: A Neglected Area of Management Policy," *Management Decision,* August 1973, p. 178.

23. Michael E. Porter, *Competitive Strategy: Techniques for Analyzing Industries and Competitors* (New York: The Free Press, 1980), p. 35.

24. Porter, p. 37.

25. Porter, pp. 38–39.

26. Al Ries and Jack Trout, *Positioning: The Battle for Your Mind* (New York: McGraw-Hill, Inc., 1981).

27. Ronald R. Gist, *Marketing and Society,* 2d ed. (Hinsdale, Ill.: The Dryden Press, 1974), p. 49.

28. Maher, p. 54.

8 Strategic Market Planning

In any corporation, strategic planning is a must. Complementing the corporate plan and giving it market relevance is the strategic marketing plan. In the turbulent world of high technology, strategic planning becomes even more critical.

As we indicated in our introduction, we are offering more of a "what to do" focus than "how to do it." Therefore, we are not presenting a discussion of the intricacies involved in the development of a corporate or marketing strategic plan. Certainly, the various aspects of planning, such as objective-setting, portfolio analysis, strategy formulation, and resource allocation, are necessary. But our primary concern is to acquaint the high-tech marketer with the importance of considering the impact of future change on his or her organization. Thus, much of this chapter is devoted to the use of scenarios as a formal way of accounting for the inevitability of change. Other planning matters are left to comprehensive strategic planning publications. (For additional reading, please consult our *Strategic Planning: Concepts and Implementation,* New York, Random House, 1984.)

Change Is the Only Constant

Fortune magazine for many years has published its annual ranking of the five hundred leading U.S. industrial corporations, as measured by sales revenues. About half the companies that were counted as members of the *Fortune* 500 in the mid-to-late 1940s are no longer on the list, which says it is easier to get on top than to stay there. Many companies, including a fair share of high-tech firms, that were not even founded until much later than the 1940s are today members of *Fortune's* prestigious listing.

Why is it that numerous industrial leaders, with supposedly the cream of the management crop as their stewards, have vanished from the *Fortune* 500? The pitfalls are really quite simple to pinpoint conceptually, but hard to avoid in practice. They are top management's rigid adherence to a *continuity assumption* and, sometimes, its *invincible ignorance* about the future. Executives in enormously successful enterprises often find it difficult to fathom that their companies will not keep on prospering eternally. They ask, "Won't there always be a market for . . .?" and "Aren't we the unconquerable leader in this market?" This continuity assumption—that management need but extrapolate the past to the future—is human and understandable. Moreover, such a premise was not too far off base until stagflation, oil cartels and embargoes, baby busts instead of booms, intense foreign competition, and consumer attitudinal changes considerably modified the business terrain. An age of discontinuity rather than continuity has best characterized the last decade or so. As a consequence, executives who kept on doing what made their companies successful in the first place usually found themselves painfully out of tenor with the times.

In major companies, the chief executive officer typically acquires the CEO's position at fifty-five years of age or thereabouts, and then serves eight to nine years until retirement. The physical and mental demands of running a large organization are too arduous to expect much more. The CEO frequently has been in the same industry all of his (most are men) working life, if not in the same company. Solely by virture of his longevity in the industry, he may have developed an incurable blind spot—an invincible ignorance—when it comes to his company's future, although there are certainly notable exceptions. He is not likely to be receptive to the notion that his company faces a diminished, maybe even bleak, future unless strategic changes of some magnitude and direction are undertaken. It is not often that a CEO radically alters the basic theme of his corporation, even when that concept is inappropriate to the future. The sizeable turnover in the *Fortune* 500 stands as a case in point. Even today, the slow- or negative-growth tobacco companies, which ostensibly have diversified into other industries, still derive the lion's share of their revenues and profits from tobacco products.

Most organizations that endeavor to install a strategic planning system to deal with change encounter significant resistance in their management ranks. Surprisingly, resistance is common even in many fast-paced, high-tech companies. It is hard for high-tech management to be concerned with planning for long-term contingencies when today is so exciting and tomorrow is so uncertain, let alone the all too vague day-after-tomorrow. In most instances, objections to strategic planning come from a lack of understanding about exactly what strategic planning is all about and what it is intended to do. Several arguments against strategic planning are recurring enough to warrant attention.

We really can't plan in our business because of uncertainty. Because of technological advancements, we don't know what next Tuesday is going to bring about, let alone three, four, or five years from now. Executives who believe this line of thought—and there are usually more than a few in any organization—err by confusing forecasting with strategic planning. Most forecasting methods are based on the premise that the future will be a continuation of the past. Using historical structural relationships between and among variables as a guide, quantitative estimates are made for gross national product, inflation, interest rates, and other factors that determine and are themselves determined by the general level of business activity. Indeed, it is correct that forecasting of this genre frequently has been wide of the mark in recent years; the popular econometric forecasting services have not done well at all. But again, forecasting is not synonymous with strategic planning. In fact, strategic planning is necessary precisely because forecasting is so imprecise, especially in the long term (more than one year hence).

No mortal is clairvoyant about the future. Consequently, planning for the future—in the singular—is not prudent. It's highly risky and invites trouble. In bona fide strategic planning, alternative scenarios depicting various plausible futures are constructed. Normally, there needs to be a worst-case scenario (if things go badly), a best-case future (if things really go well), and a most probable scenario (often an extrapolation of the recent past).

These scenarios are used to conduct sensitivity tests on key elements of the strategic plan. For example, one might ask what effect a worst-case scenario would have on the company's financial performance goals and the strategies that management has mapped to attain them. If the company immediately undertakes the ambitious capital-expansion program specified in the strategic plan, what will be the denouement if the worst-case future turns out to be reality? Will the company go bankrupt? If so, is such a gamble acceptable to management and shareholders?

Essentially, when alternative futures are used, the company has multiple strategic plans—a contingency plan for each scenario sketched. Some objectives and strategies will be scenario-neutral, achievable no matter what scenario turns out to be closest to correct. For instance, a company's objective of getting out of the software business via the strategy of selling its software operations could be irrevocable, irrespective of which scenario comes about. Other objectives and strategies will be scenario-sensitive—only achievable or undertaken or advisable if certain future events result. Nearly all financial and market performance indicators vary across scenarios.

When decisions are made, they are made with assumptions in mind. These assumptions should be documented and periodically referred to in order to see if the assumptions are still valid.

We don't have time to plan. It takes too much attention away from the day-to-day operations—and day-to-day decisions are what show up on the bottom line. A chief executive officer of a major company recalled that this was the general attitude of management in his company when he arrived to become CEO. He remembered the selling job he had in persuading reluctant executives that, if they did not devote time and energy to strategic planning, someday there wouldn't be any day-to-day operational decisions to make.

The company would lose its competitive edge in an external environment that passed it by. Some of the incumbent managers were persuaded by the CEO's logic, others were not, and either moved on or were replaced. It simply does not suffice for top management alone to be convinced of the benefits of strategic planning; to be effective, the proper strategic planning mentality must pervade the entire management cadre.

Rockwell International, in response to ominous changes in the marketplace and a declining financial situation, made a commitment to marketing in 1979. With the aid of an outside consultant, strategic planning was used to "home in" on Rockwell's customers' needs and preferences. Culture committees were set up to ascertain whether Rockwell's corporate culture—its traditions, values, and beliefs—was in line with the new strategies. The committees studied the culture to learn if it was facilitating or impeding the implementation of the new strategies.[1]

If you focus too much on planning, nothing ever gets done. Planning is for dreamers, not for doers. The reverse is true. The value of strategic planning is that it facilitates proactive decision making today; it aids management in making current decisions. A common misconception is that strategic planning locks in future decisions. It does not. What strategic planning does is force a company's management to consider what effects the contingencies (embodied in the alternative scenarios) of the future might have on the firm and what management can do *now* to poise the company to seize on whatever opportunities result and to survive adversities. Because of its action-oriented emphasis on current decision making, strategic planning is rightfully the bailiwick of doers, not dreamers. This is exactly why planning requires the continual, wholehearted involvement of line managers, including, most of all, those in top management. Strategic planning is too vital to leave primarily to staff planners. One sure way to guarantee that strategic planning will not be accepted in an organization, and thus will be ineffectual, is for top management to avoid active participation in the planning process.

Dwight D. Eisenhower has been quoted as having said something like, "Plans are nothing, planning is everything." The general was correct; it is the process of planning rather than the actual written plan itself that is more important. Effective planning is a dialectic process (the Hegelian evolvement of thesis, antithesis, and synthesis). Or, stated differently, a dialectic is a method of searching for the truth by disclosing the contradictions in another's arguments and overcoming them. In this regard, the adage that, in problem solving, "two heads are usually better than one" is salient. Through a kind of never-ending dialectic among line managers, aided by staff planners, assumptions, opinions, and biases about the future direction of the company are tested for logic, consistency, and feasibility. Usually, a consensus is forged that is reflected in a widely accepted formal strategic planning document, thereby assuring better implementation.

Plans that are the end results of participatory planning of this sort have the correct correspondence between decision making on the one hand, and accountability for those decisions on the other. Line executives who immerse themselves in the strategic planning process are shaping the future destiny of their company—a destiny that they alone will be held responsible for.

Composing and Using Alternative Scenarios

Executives inexperienced in integrating alternative scenarios into the strategic planning process tend to err in one important respect while they are learning. They are susceptible to writing and using scenarios that are too narrow—too company specific or too industry specific.

Workable and useful scenarios do not discuss company objectives, strategies, and resources *at all;* rather, the purpose of constructing scenarios is to pose plausible future environments for business for which corporate strategic responses are formulated. Good scenarios do, however, discuss the industry of which a company is a part. But scenarios should not be so industry-focused that they become provincial and myopic. There is the temptation for management to concentrate on one environmental influence that has historically or recently weighed heavily in determining an industry's state of economic health. A high-tech firm is susceptible to overemphasizing technological factors within its industry at the expense of other environmental influences that impinge on industry performance. Influences such as the national economic situation, political developments, and legal and regulatory trends ultimately affect every industry, and therefore cannot reasonably be glossed over.

Especially in companies operating in those industries where demand is derived from demand in other industries, it is essential to broaden scenarios. An environmental influence (for example, demographics) that does not affect an industry directly most assuredly can affect it indirectly. An executive once said that he did not need to consider population projections because his company is "an industrial marketer that sells only to other businesses." He knew immediately the fallacy of his statement: Most of his firm's customers are consumer products companies whose revenues correlate closely with population growth.

Writers of alternative scenarios normally need to research and think about at least eight environmental influences. The relative amount of attention devoted to each influence will vary across industries. For example, legal and regulatory developments will be more pervasive in a biotechnology industry scenario than in a scenario for personal in-home computers. Here are the eight items that we consider fundamental to every alternative scenario.

Technological Prospects and Occurrences. A survey of research and development vice-presidents and directors, in technologically oriented companies of all sizes, revealed a predominately unfavorable opinion pertaining to the usefulness of formal technological-forecasting techniques.[2] Like economic forecasting, technological forecasting generally has not been accurate in predicting future events. Consequently, alternative technological scenario assumptions are needed if the high-tech firm is to prepare adequately for whatever the future holds. Different premises concerning technological developments within an industry need to correspond to and be incorporated into best-case, most likely, and worst-case write-ups about the future. The hazards of relying on a single technological forecast are too great to do otherwise. Management has not really prepared a high-tech company, in a strategic sense, if it is not ready to deal with a plausible worst-case technological occurrence—for instance, in the semiconductor industry, continued Japanese technological breakthroughs that put U.S. manufacturers at a growing cost/performance disadvantage.

Competitor Analysis. Any scenario needs to include assumptions about what a company's competitors might do in the way of broad strategic maneuvers. These maneuvers could be technological in nature or perhaps more along business lines, like a possible merger, acquisition, or divestiture. AT&T's spin-off of its operating companies, so that antitrust objections to its entry into new technologies and new industries would be mitigated, has far-reaching consequences for companies in those technologies and industries. AT&T's future presence in a given industry might not be "forecastable" with any degree of certainty. But its possible presence can be accounted for in an alternative scenario—no doubt, with AT&T's technical acumen and size, a worst-case scenario. Take the AT&T breakup from the perspective of Sprint or MCI; surely, it is a worst-case scenario. But from the vantage point of Northern Telecom Ltd., the breakup constitutes a best-case future because this Canadian equipment manufacturer sees opportunities for selling to former Bell affiliates.[3]

Political Situation. Person-specific hypotheses about who is going to be elected president of the United States, or installed as chairman and Communist party leader in the Soviet Union, or who will seize power in a coup in a Third World country in which a company has a presence, are not necessary. All that political portions of scenarios are intended to do is to account for conservative/moderate/liberal trends in a country (or region of the world) that affect the climate for business in general and perhaps an industry in particular. One major aerospace firm uses worldwide political scenarios to gauge how much money the U.S. government is likely to spend on high-tech defense in the five-year period covered by the scenario. Interestingly, in the most optimistic scenario (for the company, not the U.S. taxpayer and citizenry), tensions between the leaders of the United States and the USSR are described as strained.

Domestic political trends foreshadow what the economic environment will be like for business. They also indicate what legal/regulatory approach government will most likely take toward the private sector.

Economic Situation. As demonstrated in the recessionary period of the early 1980s, no industry and no company is truly recession-proof—not even those in high technology. And, since economic forecasting is so problematic, it is immensely important for any company to prepare for several economic contingencies. Gambling on a single forecast about inflation, interest rates, unemployment, government spending, and related economic variables is just that—gambling.

Legal and Regulatory Climate. Similarly, alternative scenarios are required to deal adequately with a number of plausible governmental legal/regulatory approaches to business generally and to specific industries. Historically, as with economics, these approaches have varied with whatever political persuasion is in practical control of the federal government. In addition to very general legal and regulatory developments that affect all businesses, a company's scenarios also need to touch upon industry-directed legal and regulatory portents that can damage or help the industry—for instance, domestic content laws (computers), health care cost controls (biomedical instrumentation), and revolutionary patent rulings (biotechnology).

Worldwide, the climate has shifted to more marketing regulation, as seen by recent European Community and United Nations' actions. Therefore, the issue becomes one of estimating how quickly additional regulation will occur that will impinge on marketers' efforts.

Demographics. The businesses and industries that are not directly or indirectly affected in a significant way by population changes are few. Population growth rates, legal and illegal immigration, shifts in the number of people in various age groupings, and alterations in regional balances of power due to people moving in and out, in some way impinge almost all companies and markets. Even high-tech, defense-industry companies that are not directly affected by the U.S. population growth rate are nonetheless affected by such population factors as the availability of skilled workers and technical professionals in areas where their plants are located.

Cultural and Social Patterns. Every conceivable company and industry is touched by cultural and social patterns. A society's prevailing cultural and social mores underlie its laws and government policies and also determine what consumers are willing to buy. Mores set the tolerance level for what society is willing to accept from the private sector. Computer firms will prosper in cultural settings where education is esteemed, technological advancement is generally equated with societal progress, and personal and commercial efficiency are valued. Yet, these firms must be attuned to society's concerns about proper data usage and privacy rights. Biotechnology companies are rapidly moving forward in R&D and its commercial application in the United States. These same firms could not operate at all in some countries because what they do would offend the religious, ethical, and moral sensibilities of the population majority. An industry trade journal, *Bio/Technology,* has noted in an editorial that the average man-on-the-street may see science fiction visions of "mutant monsters, clones of Hitler [and] . . . conspiracy between the wicked, the powerful, and the misguided but brilliant scientists" when thinking of biotechnology.[4]

Ecological Concerns. Judicious strategic management requires planners to look at plausible futures concerning the availability of the human resources and the raw materials, notably energy, that a company must have in order to compete. In high-tech companies, there is always the issue of whether there will be sufficient capable technical personnel. With the decline in math and science education in the United States, this issue is of real concern to American high-tech firms, particularly for what it presages about head-to-head competition with their Japanese high-tech competitors. A disaster scenario in this regard could easily be used by a company to take action that would improve its own strategic position and that of American high-tech companies in general. Already, for example, a number of prominent companies are contributing sizeable sums of money to engineering schools so that the schools will be able to hire the high-quality professors that are needed to educate more and better prepared engineers. Along the same lines, computer manufacturers and marketers are donating microcomputers to primary and secondary schools, as well as to colleges and universities. Even though these are undoubtedly acts of corporate self-interest, they will have the ultimate effect of technologically and economically benefiting society en masse.

In high-tech corporations, market conditions are so fluid that useful alternative scenarios have to be updated at least every other year, or annually if market changes dictate. Moreover, if a company does business internationally, it will have to prepare separate scenarios for the United States and for each country or region of the world in which the company has a presence.

Medi-Redi Labs

Struggling to Predict the Future

Long-range planning is not new at Medi-Redi, nor is the proverbial dilemma of trying to plan and forecast five years into the future when one can hardly guess what is coming tomorrow.

As a manufacturer of high-technology medical equipment and medical supplies, Medi-Redi is among the top ten in the industry in sales, net income, and ROI. (Over 2,000 firms compete in this area, but the top ten account for about one-half of the shipments.) Medi-Redi executives pride themselves on being able to anticipate the future. This enables them to avoid serious mistakes and seize key opportunities—at least up to now.

At a recent meeting, there was a concern expressed that maybe good luck had been a major factor in the past. When discussing the future, Julia Mount, R&D Director, summed it up by saying, "If only we had a crystal ball." Bill Evans, Marketing Manager, reminded everyone that top management had provided them with future scenarios to use as aids in planning.

A set of scenarios provides hypothetical portrayals of what the future may hold in light of political, social, legal, demographic, technical, and other trends. Each scenario paints a different view of the future and thus allows planners to evaluate various contingencies.

Bill read from one of the scenarios:

The Years of Opportunity: 1984–1988

The future is characterized by opportunities made possible by technological advancements and innovations coupled with a continued wave of worldwide health consciousness. Other positive forces come from a more permissive political attitude concerning business regulation. This trend is also reflected in legislation promoting R&D in the health care area. The economy remains uncertain, fostering fierce competition. This competitiveness can only help those who are already industry leaders, if they stay alert.

Growth at both ends of the population spectrum spurs the need for medical products. While not yet a baby boom, marriage and family are both becoming popular again. The population continues to live longer. Living longer means more demand for medical and health care products. Those over age 65 comprise more than 12 percent of the population; about 30,000,000 people. Those age 45 to 64 comprise about 23 percent of the population and nearly 40 percent of them earn $25,000 or more annually.

Not only do they have money, but they are also a narcissistic group bent on staying young and healthy. Retail sales of vitamins have topped one billion dollars.

A large portion of the population has been unable to control certain addictive behaviors related to substance abuse. Over 50 percent of the adult population smokes cigarettes. An average of eight million people smoke over a pack a day.

Acute medical conditions, such as infections, digestive problems, and injuries, continue to rise. Those limited by chronic conditions in the population have grown from 23.2 million in 1970, to 31.5 million in 1980, to 36 million at present.

In summary, more emphasis will be on health and prevention by many, and abuse of health by others.

The Social Security and Medicare systems are sound, at least to the end of the decade. Millions of senior citizens form a powerful lobby, and one of their concerns is for a long, healthy, secure life.

With probusiness candidates doing well in the upcoming national elections, doing business becomes less burdensome, less regulated, and less filled with paperwork. Advertisers also enjoy more freedom.

In addition, the trend to encourage manufacturers of drugs to conduct more research is as strong, if not stronger, than in 1982, when Public Law 97–414 was passed giving federal income tax credits for up to half the cost of clinical testing of drugs.

Longer patent protection for drugs and other chemicals is now the law. The 17-year patent life has been extended to 21 years. This extension means generics are moving more slowly into the market.

The health care industry remains relatively immune to economic fluctuations. Unemployment continues to be high (8 to 9 percent) and inflation has again begun to edge up to near double digits. Workers are less militant than in the 1970s. Employers are more democratic. Many ideas have been borrowed from the Japanese and others to increase creativity and productivity.

Interest rates are generally high, with a prime rate of 12 to 14 percent. Firms who are industry leaders have been able to support growth out of current earnings.

Total national health care expenditures, as a percentage of GNP, have moved from over 10 percent in 1970 to 13 percent at the present time.

The single most important item in the news is medical technology. In just a few years, the artificial heart seems as old-fashioned as the first monster computers. Not only for the heart, but for other body replacement parts, technology continues to move at a rapid rate. For the average citizen, "future shock" continues.

Pharmaceuticals are being tailored to meet nearly every medical need. Just like breakfast cereals and cars, there is a pill for every purpose and purse.

Mental depression, estimated to have affected 35 million Americans (10 million severely) at the beginning of the decade, is now in 80 percent of the cases being managed successfully with chemical treatment.

On the horizon are solutions to solving the world's food problems. Artificial foodstuffs, supplements, and even a handful of pills have sustained life in monkeys in laboratory tests.

With the construction start of the first outer space manufacturing center planned by the U.S. government for 1990, manufacturers are faced with major long-run choices about plant, equipment, and manufacturing methods. The U.S. Government Printing Office has just issued the guidelines and applications. One firm from each major manufacturing industry will be accepted for the experimental project.

While some firms feel robotics offer the only solution to rising manufacturing costs, others have held fast to their belief that it will be decades before robotics are widely used in manufacturing.

Many in the health care industry have found the best source of new markets to be overseas. What may have peaked here is just catching on in Europe, Asia, and elsewhere. Austria, for example, has continued its expansion and modernization of health care facilities with a projected investment outlay of 2.3 billion dollars by the end of the decade.

The Japanese have proved to be good partners for joint ventures because of both their technical skills and worldwide marketing experience. And, there may be a shortage of technical personnel to hire in the United States, given declines in American math and science education.

The takeover mania has spread to the health care industry.

......

When the presentation ended, someone asked, "Where do we go from here?" Julia responded, "I think we may have found our crystal ball." Bill agreed and suggested that everyone review the three scenarios sent by headquarters before their next meeting, when objective setting and strategy formulation for each scenario would begin.

Medi-Redi Labs is a composite of several companies.

The scenario-writing team should be interdisciplinary (e.g., a demographer, an engineer, a lawyer, etc.) in order to achieve balance and a sufficiently broad and comprehensive viewpoint. Those executives who are principally involved in strategic planning for a company should not be on the scenario-writing team; otherwise, there is the temptation for them to write scenarios to fit preconceived notions about which strategies are best to pursue. Nor should strategic planning executives know the relative probabilities of occurrence that the scenario-writing team has assigned to the scenarios, lest the planning executives devote inordinate attention to the most likely scenario and give short shrift to the ostensibly less probable futures.

Window on the World

The purpose of researching and writing alternative scenarios is to identify potential opportunities and threats in the environment external to the high-tech firm. The intent naturally is to enable top management to steer the company toward the opportunities and away from the threats.

Evaluating the events taking place in the outside world, however, is only part of the strategic planning task. A company additionally needs to assess its internal readiness—its strengths and weaknesses—in terms of human and financial resources and degrees of competence in marketing, operations, and the technical side of the business. Still, there is much more to strategic management and planning. The element that enables management to seize upon external opportunities is timing. The adage that "timing is everything" overstates the case, but nevertheless gets at the right idea. Timing matters a great deal in strategic planning and particularly in the turbulent world of high technology, where things happen so fast. Propitious timing is what gives strategic planning its name.

It is not enough for management to identify potential opportunities via the process of researching and developing alternative scenarios. It is also not enough to have the company internally strong in R&D, marketing, engineering, operations, finance, and human resources. For best results, there must be a matching of external opportunity with internal capability at the right time. If an external opportunity becomes available and the company is unable to exploit it for some internal reason, this is not strategic management. Or, if the company is internally strong, but opportunity has passed it by, this is not strategic management. Strategic management is derserving of the name only when the company's executives have demonstrated the ability to align external opportunity with internal capability.

Another way to view the crucial need to match external market opportunity with internal corporate capability at the right time—the basis of strategic management—is through the concept of the "strategic window."[5] Visualize a window with the company's top management watching through it on one side, ever alert and anticipating opportunities (unmet needs and wants) that appear and pass by on the other side . . . and quickly pass by in high technology. Provided that the company has the internal competence and resources to capitalize on the opportunity, the strategic window is open and management can seize the chance. If the timing is not right, the window is closed and the opportunity is lost—probably to competitors.

The saying that goes "Always a day late and a dollar short" describes most of us and most companies. In life and in corporations, strategic managers in the true sense are not numerous. Recall that about half the *Fortune* 500 turned over in the period from just after World War II to the early 1970s. These companies could well have afforded to hire the very best strategists available. Yet, most of them went with financial planners to fit their concept of strategic planning—the very group that is most likely to extrapolate the past, via numbers, to predict the future. Top corporate managers persuaded of their industries' economic immortality, reinforced by their financial planners, were blind-sided by a world of changes that they neither expected nor understood.

Strategic market planning within the framework of alternative scenarios can aid executives markedly in improving timing aspects of their long-range decisions. The alternative scenario approach to strategy formulation encourages managers to think about various contingencies and to rehearse and to articulate to colleagues in written plans how they would handle each simulated future. Alternative scenarios require executives to go beyond the strategically precarious but comfortable habit of getting the company ready for a tomorrow that is but a continuation of the past. The "sin of the generals"—the assumption that the next war will be fought like the last war—is why it is easier to get a corporation on top than to keep it there. This observation is more true of high-tech firms than most others. Early on, new technologies carry the high-tech firm forward. Later, as its markets mature, the company and its management become more a part of the establishment, more complacent, more prey to the "sin of the generals."

The alternative-futures overlay on strategic planning requires managers to take preparatory steps now to cope with future uncertainties. (Again, to reiterate an essential point, strategic planning focuses on making better current decisions rather than on locking in future decisions.) This process of providing tentative answers, in written plans, to plausible "what if" questions serves to temper the initial disorganization and perhaps, in some cases, immobilizing shock when unexpected events—both favorable and unfavorable—confront top management. Consequently, their company is less vulnerable to events. Executives become more adept at seizing opportunities in a strategically timely manner and in coping with adversities.

Putting Scenarios into Practice

A healthy amount of freewheeling creativity is needed in the strategic planning process. However, creative thinking needs the imposition of structure if it is to be channeled constructively toward desired ends. A planning sequence of creativity followed by structure is conceptually and procedurally akin to the process seen time and again when an inventive, entrepreneurial type of person comes up with a new product, service, or venture, but shortly thereafter sells it or turns it over to the stewardship of professional managers. This sequence allows the creativity embodied in the innovation to be harnessed and shepherded toward growth and profit objectives.

Too frequently, organizational planning is strong on structure and weak on entrepreneurial creativity—which is not surprising since, like most other people, corporate executives are more comfortable with the usual than they are with change. By education, training, occupation, and proclivity, corporate staff executives are not normally entrepreneurial types who initiate ventures, nor is creativity usually a job prerequisite for them. Analytical skills are more typically their forte. Most corporate planners are rather atypical citizens, in that there is a certain insularity to their lives. Professionally, educationally, and socially, they tend to be isolated from the workaday world of mainstream America. Thus, they become somewhat debilitated in their ability to discern early on and to interpret correctly the fundamental changes occurring in society. This limitation can lead to serious errors of judgment.

The strategic-planning approach to change we prefer strikes a balance between creativity and structure; it incorporates provisions for stimulating plenty of creative thinking while also providing the structure needed to guide such creativity toward goals. It fosters considerable participation in planning by policy-making executives, staff planners, high-level operating managers, and, significantly, nonmanagerial kinds of people who are far more representative of society—and what is going on in it—than any of the aforementioned groups.

The ensuing discussion uses an example of a company organized on a divisional basis. The terminology *corporate level* refers to the parent company and *division* designates an operating, profit-center subsidiary of the parent. In practice, a participative approach to planning for change can be adopted and employed by anything from a giant conglomerate to a single small company. One procedure is described here.

Initially, corporate policy-makers, with the aid of a planning staff, write up descriptive, explanatory scenarios for each contingency and trend that they want divisional management to consider. This step forces top corporate management to focus and set down their thinking about the future. The written scenarios also serve to set parameters and to define the problems at hand for divisional managers.

Second, each division's ranking executive (general manager or president, for example) appoints a planning-project team comprised of herself or himself and the division's ranking managers from the major functional areas, such as manufacturing, marketing, finance, legal, and personnel.

Third, corporate level provides each divisional planning-project team with written contingency and trend scenarios.

Fourth, members of each division's planning-project team are asked to develop—individually—a list of conceivable opportunities, problems, and solutions that each trend may present for the division and its products and services in three or five or ten years. Team members are asked to consult widely with employees in their functional areas (including, if feasible, a sampling of clerical and blue-collar workers), as well as to gather information and to think on their own about the possible effects of each trend.

All participants are asked to "telescope" and "explode" their thinking, that is, to think in the future tense and to examine how the contingencies may affect not only the division, but also its customers and, if applicable, its customers' customers—an important consideration for any industrial division whose fortunes wax and wane with derived demand. Small group meetings, focus groups, and brainstorming sessions are useful in gathering the types of in-depth, uninhibited input sought. It is our experience that executive retreats are beneficial in this regard; they allow managers to get away from the telephone and day-to-day pressures that tend to militate against free, innovative, and insightful long-range thinking.

Finally, the members of each division planning-project team meet to discuss, debate, and then sort out the most plausible opportunities, problems, and solutions from those proposed. Results of these deliberations are committed to writing, complete with supporting rationale, and submitted to the corporate planning staff.

The entire process might take possibly two months to complete. Each division's written report discussing how trends are likely to affect it can be used by the division's top management to formulate the operating unit's long-range contingency plans (one for each major scenario). Similarly, at the corporate level, the reports prepared by the divisions can be used by policy-makers and planners to identify potentially high-, medium-, and low-return divisions, as measured against corporate growth and profit objectives. Consequently, top corporate management should be better able to make informed internal investment and divestment decisions and, if need be, to develop and pursue external acquisition and merger strategies appropriate to the attainment of corporate goals.

Bureaucracy and Formal Planning Systems

Experience makes clear that there is a tendency for management to install excessive bureaucracy and too many formal procedures in the formative stages of organizing a company's strategic planning system. In the most successful companies the operation is streamlined and pruned as time passes. Companies that organize and plan around strategic business units (SBUs) often begin with too many SBUs, but gradually consolidate and pare the cumbersome number of SBUs as experience in practice dictates. (Strategic business units are relatively self-contained businesses within a corporation which compete in the marketplace. SBU managers are responsible for developing and implementing their own business strategies, subject to corporate approval.) We believe that this initial erring on the side of too much bureaucracy is the wrong tack.

Bureaucracy and formal planning procedures are required if the strategic planning task is to get done. Carried too far, however, bureaucracy and formality inhibit a company's management from taking the quick actions required to capitalize on market opportunities and to weather setbacks. Management's overadherence to bureaucratic arrangements, organization charts, and formal procedures makes a company ponderous and lackadaisical, dampens creativity, and encourages a close-to-the-vest executive style. These are significant competitive disadvantages to all firms and, sometimes, the decisive "kiss of death" to companies that operate in the fast-moving world of high technology.

Thomas Peters and Robert Waterman, Jr., authors of the best-selling book *In Search of Excellence,* share their findings as to why excellent American companies stand out. These best-managed corporations have traits in common:

> Informality and open communications. The president of Walt Disney Productions wears a name tag bearing only a first name and IBM's chief executive answers employees' complaints.
>
> Structural Flexibility. Top management shifts human and financial resources quickly within the company in order to respond in a timely fashion to changes in the external environment, even in a giant company like IBM.
>
> Adhocracy. Task forces are short-lived by design and staffed with busy senior executives who want to get the job done and get on to other matters. Consequently, task forces and committees are formed for only important issues.
>
> Risk-Taking. Experimentation and innovation are expected.[6]

In sum, Peters and Waterman found among the best-managed companies: (1) an action-orientation that rewards ad hoc informality and open communications in problem solving, in lieu of bureaucratic behavior; (2) a well-defined focus outward toward the marketplace instead of inward toward the company; (3) a clear sense of direction about where the company intends to go; and (4) a preference for simplicity (a former Proctor and Gamble president demanded one-page memos from his subordinates).

We have observed that management in prospering high-technology firms embody and personify these traits. A robotics company's top executive comes to mind. He is a "sixty-ish" ball of energy and enthusiasm, who involves himself integrally in strategic marketing decisions. The question is, of course, can such enthusiasm be maintained as the company matures and grows?

Terrence Deal and Allan Kennedy are the authors of the book *Corporate Cultures.* Their research has led them to conclude that midlife crises are not at all unusual for high-tech firms. Some companies, as they mature and grow, are unable to sustain the culture and direction that carried them to success. According to a *Wall Street Journal* article, Polaroid, although still comfortably profitable, is such a company.[7] Sales are shrinking, turnover of talented employees is on the rise and, reportedly, the corporate esprit de corps is not what it once was. It may be a commentary that Polaroid's chief executive refused to be interviewed for the *Wall Street Journal* article about his company's problems and he also asked his subordinates to decline interviews. This purportedly reserved and cautious approach stands in contrast to the openness that Peters and Waterman found in America's best-managed companies.

In our own discussions with executives in successful high-tech firms, we encountered refreshing forthrightness—not about trade secrets, naturally, but certainly about corporate objectives, strategies, and operations, including problems. This candor need not inevitably erode as a high-tech company grows larger. IBM is a massive corporation that has been able to foster and retain a "can do" culture.

The linchpins underlying organizational excellence identified by Peters and Waterman are insightful, but not really new or surprising. Such managerial principles as simplicity, open communications, structural flexibility, and the minimization of red tape, are eternal verities that have long been recognized as characteristic of successful managers and effective organizations.

If these are the elements of winning management, why don't more executives and companies emulate them? The lasting influence of the military form of organization and administration is perhaps the major reason. Some managers are detached and reserved by nature, and prefer to operate behind and through a veil of subordinates, procedures, and formality. Many executives equate a staid, highly structured work environment with the proper business atmosphere. And, in all candor, long-term task forces and committees to study problems are sometimes used as a means to delay unpleasant decisions. Then, too, an organization can quite unintentionally become a victim of its own success. Bureaucracy is unwittingly allowed to build up as the organization grows and prospers. This executive needs an assistant, that executive needs an assistant, and so on, until soon multiple organizational levels and elaborate organizational charts and reporting procedures begin to take their toll and to stifle the flexibility and creativeness that the company once had.

We have observed that the newer corporations, such as many of those found in the Southwest, tend to have a smaller middle management cadre. This leanness permits them to make quicker judgments than the more traditional Midwest firms.

Too much bureaucracy is especially dysfunctional in the strategic planning function. Line managers, who should be heavily involved in developing plans that they alone will have to implement and be held accountable for, tend to be eased aside and supplanted by a group of full-time staff planners.

How does top management go about determining the right amount of planning bureaucracy? Surely some staff planners are essential. They are needed to provide invaluable support to line executives. But how many is enough? How does management hold down bureaucratic excess, yet not be so understaffed that executives are spread too thin?

Approximations of the optimum amount of bureaucracy can be achieved through a racket or incremental approach to developing the planning organization. Incrementalism enables management to add organizational levels and staff gradually, and only after a clear need is demonstrated. The alternative is a sudden planning approach, whereby an organizationally pervasive system is imposed virtually all at once. In addition, the planning organization needs to be reviewed periodically to determine whether it has gotten too complex. General Motors' CEO said that one silver lining in the otherwise disastrous automobile slump of the early 1980s was that the GM organization had been trimmed to "mean and lean" proportions, never to be built back up to prerecession levels.

Linking Market Strategies to Technical Strategies

The purpose of strategic planning is to provide a guiding theme for where top management intends to take a company in the long term (3, 4, 5 years hence) and a blueprint mapping how it intends to do so. A high-minded-sounding philosophical statement that says nothing specific will not get the job done. Neither will a plan that couches goals and objectives primarily in financial terms.

The corporate theme needs to be clearly identifiable from a written market/technology-oriented business definition. Statements like "We are a high-technology medical instrumentation company" are nebulous and thus give only a modicum of direction to marketing management and to research and development. Such statements may even give misdirection. A comprehensive business definition addresses four issues:

1. What customers (market segments) the company intends to compete for?
2. Which customer needs and desires the company intends to fulfill?
3. What technologies will the company use to fulfill these needs and wants?
4. What will be the geographical scope of the company's efforts—regional, national, or international?

For a high-tech company, the question of what technologies the company will use to fulfill these needs and wants is especially crucial. How it is answered tells R&D whether management intends to stay mostly with existing technology or to seek something new and innovative, perhaps even revolutionary. Importantly, this answer also tells R&D and marketing how much risk top management is willing to tolerate in new product/market development and, therefore, places the risk-taking decision on top management and the board of directors, where it belongs.

A strategic plan also needs clearly identifiable and interrelated corporate goals and objectives, which, collectively, designate the company's mission. The use of the word mission in a strategic context comes from the military—for example, "The mission of this army division is to launch a broad-scaled combat operation that will accomplish . . . [a series of interrelated objectives]." When adapted to business, a mission states objectives in terms of markets and financial performance targets sought by management and the board of directors.

If the high-tech company is to be successful in achieving its mission, there must be close linkage between its business strategies and its technical strategies. Were the R&D function allowed to go its own way, then only through happenstance would there be a correspondence between profitable business and technical opportunities.

Stated negatively, there are several ways for management to make certain that there will be a linkage problem. They can be vague in addressing the corporate business definition, so that it is unclear as to what market segments, what customer needs and wants, what technologies, and what geographical areas are to be focused on. Or, management can state corporate goals in the mission statement mostly in financial terms. If, for example, the strategic plan says that the goals are to achieve a certain after-tax return on investment, to maintain a minimum current ratio, to be first or second in market share in every market in which the company competes, and so on, then many R&D strategies could conceivably achieve the financial goals. The R&D people are left with no guidance, and their efforts may or may not coincide with present or future market opportunities.

To be sure, financial goals are important and must be an integral part of a strategic plan. Yet, one should not lose sight of the fact that financial achievement is but a by-product of successful corporate performance in the marketplace. Rockwell International, in response to its deteriorating financial performance, decided to focus on its customers' needs and preferences. Rockwell put the horse and the cart in the right order. The most successful companies—those which consistently outperform the competition—are always pointed outward toward the marketplace and, consequently, have a keen sense of direction about where they need to go to attain their financial objectives.

Still another way for top management to create a technical-market linkage problem is for them to fail to communicate adequately their definition of the business and their mission for the company. Setting the corporate agenda for the long term is rightfully top management's responsibility and prerogative. However, R&D must know and understand the agenda if it is to do its part to make the agenda a fait accompli.

Edward Weil and Robert Cangemi's survey of corporate R&D vice-presidents in high-tech U.S. corporations revealed several causes of what Weil and Cangemi confirmed to be a substantial problem of linking strategic planning with research.[8] Especially illuminating is the sharp contrast between survey respondents who reported no linkage problem and respondents who perceived a severe linkage problem.

The R&D executives who saw no linkage difficulties in their organizations indicated company cultures with plenty of collaboration between top-echelon corporate management and R&D, much flexibility, and frequent two-way communications. More specifically, several organizational arrangements and techniques were characteristic:

> A combination of top-down and bottom-up research program generation is used. Top corporate management points out new areas in which R&D programs should be started and to which researchers or task forces are to be assigned. Afterwards, the actual long-range research program is generated from the bottom up.
>
> The corporate planning group, marketing management, and operating executives, all participate in decisions on the long-range research program.
>
> Staff representatives of top management regularly attend lower-level research planning and review meetings; and informal contacts between top managers, strategic planners, and researchers are encouraged.
>
> Researchers are given a free rein and indeed are encouraged to seek out and explore unsatisfied needs of the marketplace.

A quite different picture emerges from research managers in companies with troublesome linkage problems. These executives see their function impeded by a lack of strategy from corporate top management and by pressures that are counterproductive to achievement of effective long-term R&D programs. In particular:

> Corporate objectives are too general to guide research with any precision.
>
> Corporate goals are described in jargon not helpful to research.
>
> There is a lack of systematic means for research to determine corporate needs.
>
> There is also a lack of effective ways for research to discern long-term market needs and wants.
>
> Corporate management seems to be opposed to high-risk ventures, which limits corporate research possibilities.

Strategic Thinking at the Societal Level

An idea that is being popularized and widely accepted in the United States is that of an American postindustrial society: America is rapidly evolving from an industrial-manufacturing country to one specializing in knowledge. Members of a high-tech-pushing faction in the U.S. Congress are even known as Atari Democrats.

In our judgment, this perspective, if accepted as inevitable and as gospel, is unwise on two counts. First, it all but concedes a large chunk of U.S. basic manufacturing to foreign countries. Second, it naively implies, with elitist overtones, that the U.S. work force is going to supply the brains—the information and knowledge—and that foreign workers will provide the brawn.

Rather than simply giving up much of this country's industrial base, a better societal strategy would be to make the basic manufacturing industries the most competitive in the world via the application of high technology—for example, via robotics and computer-aided manufacturing. After all, high technology is a means to an end, not an end in itself.

Moreover, all the brains and knowledge do not reside in America. Other industrialized nations are making gains in the commercial application of physics (West Germany, France, Italy, Japan); computing (Japan); and chemistry and biology (France, Japan). In a worldwide market, the United States cannot reasonably depend on maintaining a decided competitive advantage over these countries in knowledge industries. Strategically, the United States needs a more balanced economic approach in which high technology, on the one hand, and lower-tech manufacturing and service industries on the other, are treated as synergy-producing competitive complements to one another, instead of as either/or substitutes.

Marketing's role in achieving more economic balance in America is a key one. In fact, it is essential. Within U.S. companies, the marketing function is closest to the marketplace, where innovations from the laboratories must ultimately prove their merit. When more companies learn to do a better job of aligning what marketers conjecture and know about market opportunity with what research thinks possible in technical opportunity, marked economic progress in the United States, domestically and internationally, will be manifest. Through additional experience and experimentation, marketers can keep refining and creating techniques and organizational structures for mitigating the high risks intrinsic in the process of commercializing technological breakthroughs. A technologically oriented nation like the United States requires the very best in the way of scientists, mathematicians, and support for research and development. But a high-tech nation cannot remain a technological leader for very long without the very best in high-tech marketing also.

Notes

1. ''Marketing Approach Means Survival for Reeling Rockwell,'' *Marketing News,* May 27, 1983, p. 15.
2. Edward D. Weil and Robert R. Cangemi, ''Linking Long-Range Research to Strategic Planning,'' *Research Management,* May-June 1983, pp. 32–39.
3. Peggy Berkowitz, ''Northern Telecom Bids To Sell More Switches As AT&T Breaks Up,'' *Wall Street Journal,* December 30, 1983, p. 1.
4. Christopher Edwards, ''Consumers Must Be Prepared For Biotech Products,'' *Bio/Technology,* April 1983, pp. 137, 211.
5. Derek F. Abell, ''Strategic Windows,'' *Journal of Marketing,* July 1978, pp. 21–26.
6. Thomas J. Peters and Robert H. Waterman, Jr., ''A Bias for Action,'' *The Best of Business,* Spring 1983, pp. 6–12.
7. William M. Bulkeley, ''Losing its Flash—As Polaroid Matures Some Lament a Decline in Creative Excitement,'' *Wall Street Journal,* May 10, 1983, pp. 1, 20.
8. Weil and Cangemi.

Appendix

How Top Executives in High-Technology Companies See Themselves—Their Behavioral Characteristics and Motivations
(from chapter 1, page 9)

In each of the following items there are two words or phrases that are polar opposites which are separated by a seven-space scale. Please look at each pair of words and then place a check mark (✓) on the scale in the place that you believe best describes you as an individual. Please record your first impressions.

Traditionalist	_ _ _ _ _ _ _	Avant Garde
Nonliteral Thinker	_ _ _ _ _ _ _	Literal Thinker
Nonlinear Thinker	_ _ _ _ _ _ _	Linear Thinker
Motivated by $	_ _ _ _ _ _ _	Not Motivated by $
Tortoise	_ _ _ _ _ _ _	Hare
Gambler	_ _ _ _ _ _ _	Risk Averse
Nonintuitive	_ _ _ _ _ _ _	Intuitive
Entrepreneurial	_ _ _ _ _ _ _	Nonentrepreneurial
Inferential Thinker	_ _ _ _ _ _ _	Noninferential Thinker
High Associative Skills	_ _ _ _ _ _ _	Low Associative Skills
Illogical	_ _ _ _ _ _ _	Logical
Analytical	_ _ _ _ _ _ _	Nonanalytical
Noncreative	_ _ _ _ _ _ _	Creative
Word/Symbol Oriented	_ _ _ _ _ _ _	Pattern Oriented
Nontechnology Buff	_ _ _ _ _ _ _	Technology Buff
Promote/Welcome Change	_ _ _ _ _ _ _	Resist Change
Extrapolative Thinker	_ _ _ _ _ _ _	Nonextrapolative Thinker
Low Ambiguity Tolerance	_ _ _ _ _ _ _	High Ambiguity Tolerance
Introvert	_ _ _ _ _ _ _	Extrovert

Index

About the Authors

William L. Shanklin has been involved in consulting and executive development with organizations ranging from *Fortune 500* companies to small businesses and was employed in industrial marketing by the Georgia-Pacific Corporation. His associations have included such corporations as Digital Equipment, General Electric, and Goodyear Tire and Rubber. He is a board member of Carrollton Graphics, Inc. Shanklin has published on a wide range of business and marketing topics. His articles have appeared in *Business Horizons, Director & Boards, Journal of Advertising, Research Management, Harvard Business Review,* and many other practitioner-oriented periodicals. He is a frequent speaker to business groups. Shanklin is professor of marketing in the Graduate School of Management at Kent State University in Ohio. His doctorate is in business administration from the University of Maryland. He and John K. Ryans, Jr. are also the coauthors of the book *Strategic Planning: Concepts and Implementation.*

John K. Ryans, Jr. is a prolific writer on a wide range of domestic and international business topics. Among his most recent books is the *Management of International Advertising,* coauthored with Dean Pebbles of Goodyear International. Ryans has published articles in the *Journal of Marketing Research, Business Marketing,* the *Journal of Advertising Research, Harvard Business Review,* and numerous other journals. He is the coeditor of a new journal (*International Marketing Review*) and often speaks to business groups. During his career, Ryans has been an active consultant, numbering many U.S. and foreign companies, including Novo Industri, Xerox, and McCann-Erickson, among his clients. A doctor of business administration graduate from Indiana University, he has been a visiting professor at Columbia University and the University of Houston. Currently, Ryans is a professor of marketing and international business of Kent State University and a frequent program leader at the World Trade Institute in New York.